80° 70° 60° 50°

ER SOUND

Bylot Is

BAFFIN BAY

B A F F I N

Igloolik

I S L A N D

Melville
Peninsula

DAVIS STRAIT

GREENLAND

Clearwater
Fiord

Pangnirtung

Kerkerten Is

Cumberland Sd

Blacklead Is

Bay

Maliksitaq

Lyon Inlet
Gore Bay

FOXE BASIN

FOXE CHANNEL

Foxe Peninsula

Tikkuut

Iqaluit

C Haven
Cyrus Field Bay

60°

Frobisher Bay

Pt

Southampton
Island

Mill Is

Cape Dorset

Akuliak

Lake Harbour

Coral Harbour

Spicer's Is

Salisbury Is

Big Is
North Bluff

North Bay
Upper
Savage Is

Resolution Is

illerton

Nottingham Is

HUDSON STRAIT

Coats Is

Mansel Is

Ungava Peninsula

Ottawa Is

QUEBEC

SON BAY

WHEN THE WHALERS WERE UP NORTH

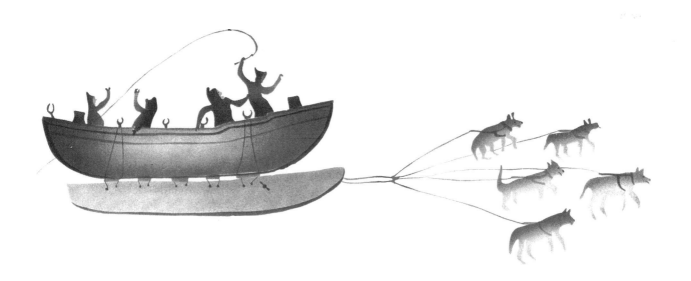

Nowyook Remembers. Nicodemus Nowyook, Pangnirtung.
Stencil, 1985.

When the Whalers Were UP NORTH

INUIT MEMORIES FROM THE EASTERN ARCTIC

Dorothy Harley Eber

David R. Godine, Publisher • BOSTON

First U.S. edition
published in 1989 by
David R. Godine, Publisher, Inc.
Horticultural Hall
300 Massachusetts Avenue
Boston, Massachusetts 02115

Library of Congress Catalogue Card Number: 89-45516
ISBN: 0-87923-818-6

First U.S. Edition
Printed in Canada on acid-free paper

Design by Ross Mah Design Associates

In different form, sections of chapter 10 appeared in "Bringing the Captain Back to the Bay," *Natural History* (Jan. 1985). Short excerpts from chapters 3 and 6 have also appeared in "Glimpses of Seekooseelak," in *Cape Dorset*, the 1980 exhibition catalogue of the Winnipeg Art Gallery, and in "Remembering Pitseolak Ashoona," in *Arts and Culture of the North* (New York, Fall 1983).

Inuit art reproduced with permission of the West Baffin Eskimo Co-operative, the Pangnirtung Eskimo Co-operative, the Sanavik Co-operative of Baker Lake, and the estate of Jessie Oonark.

CONTENTS PAGE ILLUSTRATION:
Whale Hunt. Nicodemus Nowyook, Pangnirtung, Stonecut, 1973.

CHAPTER OPENING ILLUSTRATION:
Whale Hunting. Nicodemus Nowyook, Pangnirtung, Stonecut, 1975.

CONTENTS

When the Whalers Were Up North

Without the interest and co-operation of the descendants of the Inuit whalers this book would not exist, and my first thanks go to those who so generously passed on the stories and information found in these pages. I also want to thank the many interpreters who worked with us for their major contribution.

In addition, many individuals and institutions in Canada, the United States, England and Scotland helped to make this book a reality.

I owe a particular debt of gratitude to Merloyd Lawrence, whose thoughtful advice helped me to find a framework in which to present the oral history around which this book is built. I am also deeply indebted to the scholars in the field. Indispensable during the research and writing of this book have been the splendid texts of W. Gillies Ross, Canada's premier historian of Arctic whaling. I have also benefited from the helpful suggestions of Philip F. Purrington, senior curator of The Whaling Museum, New Bedford, Massachusetts, and author of elegantly written articles in the *Bulletin*, who gave me encouragement and assistance at the outset of this project, and from the publications and counsel of Philip Goldring, Gavin White, Daniel Francis, and Fred Calabretta. For assistance in regard to aspects of Scottish whaling I must

Spring Floe Whaling. Tommy Nuvaqirq, Pangnirtung. Stencil, 1977. The artist shows Cumberland Sound whalers cutting in their kill. The whalers put spikes on their boots, to climb on the huge mammal.

thank David Henderson, assistant keeper, Natural History, Dundee Museums and Art Galleries.

For photographic advice I am much indebted as with previous publications to Stanley Triggs, curator of the Notman Photographic Archives, McCord Museum of Canadian History, Montreal. For the processing of some particularly difficult negatives I would like to thank the archives' technician Tom Humphry. I would also like to thank Patrick B. O'Neill of Mount Saint Vincent University, Halifax, and Jonathan King and his staff in the Ethnology Department of the British Museum, London, for assistance in researching the important Geraldine Moodie collection of photographs. Much help was also received from the National Archives, Ottawa, and, in connection with both photographs and documents, from The Whaling Museum, New Bedford, Massachusetts, from the Kendall Whaling Museum, Sharon, Massachusetts, and from Mystic Seaport Museum, Mystic, Connecticut, as well as other archives in Canada and the United States. The origins of all photographs and documents are indicated in the text or reference notes. Inuit artwork is published with the permission of the Inuit co-operatives and I am grateful for the help of the Canadian Eskimo Arts Council. I would like to thank the Innuit Gallery of Eskimo Art, Toronto, and Indian and Northern Affairs Canada for lending me transparencies and slides of the artists' work.

I would like to thank Jane Aupaluktuq and Veronica Curley, who directed the Interpreters' Corps in Rankin Inlet at the times of my visits, for assistance in obtaining expert interpretations. I am indebted to Makivik Corporation, which kindly permitted early drafts of the manuscript to be prepared on their computer. This task was undertaken by Hugette Milburg Topy. For other timely assistance I must thank David Audlakiak, Deborah Evaluarjuk, Bernadette Driscoll, Melvin Weigel, Marie Tessier-Lavigne, Toby Morantz, Sandra Barz, Barbara Lipton, Alan Fagan, Jean Bruce, Barrie Kinnes, Robin McGrath, Randall Reeves, Bill Kemp, David Zimmerly, George Swinton, Joan Eldridge, Mr and Mrs Austin Murray, Norah Hague, Terry Ryan, Doris and William C. Spicer, John Spicer, Doris Pulaski, Marie Bouchard, Frank Kosa, Mary Lee Powell, Jennifer Murn, and Lesley Andrassy.

I wish to express my appreciation to the Urgent Ethnology Programme of the Canadian Museum of Civilization, which retained me to collect reminiscences of the whalers, for their generosity in permitting me to publish the material, and to the Explorations Program of the Canada Council and the Multiculturalism Program of the Department of the Secretary of State of Canada for their support. I would also like to thank Nunatta Sunaqutangit, Iqaluit's "Museum of Things from the Land," for their invitations to visit Iqaluit (Frobisher Bay) and Pond Inlet. To Philip Cercone and Joan McGilvray and all at McGill-Queen's University Press I offer my sincerest thanks for their interest in my manuscript and their enterprise and care in bringing it to publication. My thanks also to Susan Kent Davidson for her careful and perceptive editing.

On a personal note, I would like to offer thanks for warm hospitality received while working in the North to Annie Manning, Chris Egan, Helen and Tom Webster, Ross Peyton, Susan and Jim Shirley, Teri Arngnaṅaaq, Margaret and Kevin Adams, Alan Dick, and Cheri Jagdeer. Finally, I would like to thank my husband, George F. Eber, who supported this book in many ways.

Dorothy Harley Eber
Montreal 1989

New houses at Repulse Bay.   Dorothy Harley Eber 1983.

It was a photograph that first spurred thoughts of this Arctic journey undertaken in the eleventh hour.

In 1981, in Cape Dorset on the Hudson Strait I sat with a group of older Inuit, as Canadian Eskimo prefer to be called, looking at images of the past. The photographs were mostly from the 1940s, 1950s, and 1960s, taken before the move to the settlements, while the Inuit still were campers, living the traditional Eskimo life. I had brought them up from the south because in some, I knew, my companions would recognize their former selves.

But one photograph I had brought along, I thought, simply for its curiosity value. It related to much earlier times. It had been taken in a New London, Connecticut, photography studio exactly one hundred years before, in 1881. The photograph showed an Inuit hunter – his name was Johnnibo – a woman, and a small girl, all three in skin parkas and white furry pants made from the thick, water-resistant pelts of seal pups. Extraordinary information accompanied the photograph: the family had been brought south by the famous New England whaling master John O. Spicer to participate in a Boston court case in which the litigants were battling over stolen whales. "This family was brought down south because of a court case," I began. But my companions already knew who the people in the picture were. There were cries of "Johnnibo!" and one man remarked, "This picture makes me remember the stories my grandmother told."

I was astonished to discover that Johnnibo's story was well remembered – and recounted with passion.

The incident changed the direction of my work. I began to wonder if it might be possible still to collect reminiscences that would give some Inuit perspective to the whaling days. Art, not whaling, had initially drawn me as a reporter to the Hudson Strait. When traditional camp life eroded in the 1950s – schools appeared – and 1960s – snowhouses disappeared – the Canadian Eskimo made a powerful response. They began to produce carvings, drawings, and prints which now have a following around the world. For several years I had been visiting Cape Dorset, trying with the help of the bilingual generation – young Inuit who have been to school – to record something of the lives of older artists whose work seems to give a last glimpse of a unique lifestyle. The idea of inquiring into the whaling days was a new one. I knew that in the nineteenth and early twentieth centuries American and British whalers whaled intensively in the waters of Arctic North America – off Baffin Island, in Hudson Bay and the strait, and off the western Arctic coast – but I had little idea of the extent of Inuit involvement. The quarry was the massive bowhead whale. Today, just occasionally, Cape Dorset hunters still sight a bowhead. "But the whales don't come into the inlets anymore; they're way out in the water – because of the ships."

Back in the south, during the fall and winter I visited the marvelous New England whaling museums, discovering more about Arctic whaling days, particularly whaling in the eastern Arctic, in their rich stores of logs, journals, clippings, and photographs. Among other things, the visits produced new chapters in Johnnibo's story. I discovered too that wintering whalers depended on the Inuit for food and clothing. Could they have survived, I wondered, without Inuit assistance? And I learned, as I later heard Inuit say, that it was often Inuit who caught the whales for American and Scottish vessels. "All these brave houses and flowery gardens came from the Atlantic, Pacific and Indian oceans," says Ishmael in *Moby Dick* of the whaling port of New Bedford. "One and all, they were harpooned and dragged up hither from the bottom of the sea."[1] Some of those brave displays were pulled from Arctic waters by Eskimo whalers.

These native whalers never kept journals, never told their stories. I heard it said that the Inuit account of whaling was simply lost. But remembering the response to Johnnibo's picture, I wondered if this were true.

With support from the Urgent Ethnology Programme of the Canadian Museum of Civilization and later from the Department of the Secretary of State's Multiculturalism Program, I returned north, visiting at various times most of the new small towns that now string along the coasts of Baffin Island and up the west coast of Hudson Bay, the communities where the children and grandchildren of the Inuit whalers live today. Sometimes these are raw towns, stressed by change, sometimes politically charged as a new society struggles to evolve. They are communities with their share of all the modern problems (high unemployment being the worst), but to a visitor they are exhilarating places with their extraordinary contrasts and populations that include survivors of overwhelming change and a rising new elite, the Inuit achievers who run for office, teach school, and manage airports.

To reach these outlying settlements and hamlets – Lake Harbour, Cape Dorset, Pangnirtung, and Pond Inlet on Baffin Island, and Repulse Bay, Coral Harbour, Chesterfield Inlet, Baker Lake, Whale Cove, and Eskimo Point in the Keewatin on the west coast of Hudson Bay – I travelled not by topsail schooner as the whalers might have done but by the short-takeoff planes that have opened up the North. Jumpoff points after arrival from the south are the larger regional administrative centres, Rankin Inlet in the Keewatin and, on Baffin Island, Iqaluit, formerly Frobisher Bay. (Residents opted recently for the old Inuit name for the location–"schools of fish.") These are civil servants' towns, with high-rises, considerable white populations, and federal and territorial as well as municipal offices and services.

Most new eastern Arctic communities grew up around Hudson's Bay Company posts after the move to leave the hunting camps began in the 1950s, but Iqaluit developed at the head of Frobisher Bay because Americans picked the site for an air force base during the Second World War. Rankin is the town that grew up around the mine. When the nickel mine started up in 1955, campers from throughout the Keewatin moved into the area and became miners. After the ore ran out, the mineshaft for some years still dominated the horizon, reminding the citizenry of what had drawn them there in numbers that eventually caused administrators to move in, too. Not long ago the shaft burned down; people miss its presence. "We ought to put up a monument," says Jane Aupaluktuq, a former director of the region's interpreters' corps. The shaft had symbolized recent history.

At first on my travels I worried whether, in the new North of today, my recording of Johnnibo's story had been just a happy accident. But this was not the case. Stories, family traditions, even gossip from the whaling days survived, and Inuit were generous with their information. "May I ask questions?" "It's the only way to find out." Soon, one informant declared, there would be no one alive to remember "when the whalers were up North."

In Eskimo Point on the west coast of Hudson Bay, the heart of Arctic America, I met Joe Curley, nephew and adopted son of Angutimmarik, or "Scotch Tom," one of the most famous of Inuit whaling bosses. Joe Curley's son Tagak is a modern Inuit leader who for many years has been deeply involved in Inuit politics. "Whatever comes up, we know that Tagak will be telling the government what to do," his father remarked. But Tagak laughed when, relating the conversation, I suggested he was something of a scrapper. "There are still so few who can speak out," he explained. "It is an obligation." From Joe Curley I learned many details of the whalers' routines. He confirmed for me also just how important Inuit partnership had to be once whalers began to winter over. In one interview he happened to mention how once, during a hungry winter, his family had had to care for Therkel Mathiassen and Jacob Olsen, two members of the mighty Fifth Thule Expedition to Arctic North America (1921–24).[2] "They were poor," he said. "They were poor like little orphan boys." Wintering whalers who had no Inuit helpers were poor indeed.

One interview generally led to another. Most older people, I discovered, had some information about the whaling years, but a few had a deeper knowledge. After a week of daily interviews Joe Curley kindly directed me to others with stories to tell. Some of

Introduction

those whose reminiscences appear here sailed on whaling vessels or lived in camps at winter harbours; two once lived at the old whaling stations in the Cumberland Sound. Others heard their parents' or grandparents' stories of the whalers. Three who filled the post of unofficial historian for their communities, often giving talks on the old days to children at the local school, had done their own research on the whalers, the "first qallunaat," or first white men. Two had published their own reminiscences.[3] In only one of the communities I visited were no stories collected. I had travelled inland to Baker Lake hoping to talk to Louis Tapatai (the name comes from "Starboard Eye," a sobriquet bestowed by the whalers) because his friends on the coast said he knew about whalers. But Louis Tapatai died the day before I arrived.

Of course I soon discovered in my journeys that Greenpeace and its ilk are not much admired in the North. The Greenpeace campaign against the seal hunt has hurt the Inuit economy. The Inuit say that the whales and the walrus and the seals are Inuit animals that give them their food and their livelihood. The qallunaaq – the white man – has his own animals. Joe Curley, in combative mood, remarked, "I do wonder

Anirnik in her bedroom, which she has brightened with sheets of Christmas parcel bows. "I haven't had time to take them down." Anirnik was born on the Dundee whaler Active *and in old age became one of Cape Dorset's noted graphic artists.*
Tessa Macintosh 1975.

why they would want to put a freeze across the Northwest Territories. I know if they should put a freeze on sea mammals, most of the native people in the high Arctic would not be able to provide for themselves because the sea creatures are their main source of meat. If they move against the sea mammals, there'll have to be compensation."

Sometimes after an interview a talented artist drew pictures to illustrate our conversation. Frequently I showed photographs. Archives in the United States, the British Isles, and Canada hold the work of a surprising number of amateurs – among them fur traders, ships' officers, missionaries, and whalers – who photographed the Arctic in early days. I also carried in my pack of pictures photographs by two professionals: Robert Flaherty, the pioneer documentary filmmaker of *Nanook of the North* fame, and Geraldine Moodie, who had had her own photographic studios in the south and was almost certainly the first woman to photograph the Arctic, taking remarkable studio-type portraits of the Inuit whalers and their bead-decked wives. Particularly fascinating to most people in the west-coast communities of Hudson Bay were the extraordinary photographs taken by the great American whaling master George Comer, aboard and from the *Era* and the *A.T. Gifford.* They were wonderful stimulants to memory. They stirred talk about the Inuit whaling bosses, Scotch Tom and Harry, and about the two Scottish whaling captains, the brothers Alexander and John Murray, and of course about Comer himself, whom Inuit call Angakkuq – the shaman – because he took photographs and "was able to perform wonders."

My talks were made possible by bilingual Inuit. Generally two interpreters, often young relatives of informants, worked on an interview. For my talks with Joe Curley, for instance, my principal interpreter during a week of interviews was one of his granddaughters, Madelaine Napayok Anderson. ("At school English was my favourite; I like a proper past participle.") Then, to make sure nothing had been missed and nuances were correctly understood, the interview tapes were reinterpreted by Joe Curley's daughter Rosie Aggark, at the time a member of the Northwest Territories Interpreters' Corps, who now has her own interpreting business in Eskimo Point. For several interviews Leah Nutaraq, who spent her youth at Blacklead Island, and I were lucky enough to have as our interpreter Ann Meekitjuk Hanson, whose appointment as deputy commissioner of the NWT was announced shortly afterwards. Ann Hanson and I had worked together in the past, and she interpreted on this occasion as a friendly gesture but also out of her deep interest "in history and the people who make history."

It was a surprise for me to discover that many younger Inuit were as unaware as I had been of the extent of their ancestors' involvement with the whalers. This was borne home by a conversation I had in one Baffin Island community that, like many, had recently acquired hamlet status. Here I met Ainiak Korgaq, a young government employee whose job was to give instruction on parliamentary procedures. Rather thoughtfully he remarked, "But first I have to have a quorum." Fine spring weather had taken the newly elected councillors out of the settlement on their skidoos to hunt caribou and birds. He told me that his grandfather had gone to Scotland on a whaler. "He died before I was born, and I never knew much about him. In history at school we learned about Sir Martin Frobisher coming into Frobisher Bay, but nothing much seemed to happen after that until the arrival of the Hudson's Bay Company. What went on in our history in the years in between?"

It is a fact that so much has happened since the whalers left (around the time of the First World War) that their once-dominating presence is often overlooked. It is usually the Hudson's Bay Company fur traders, who arrived after the whalers' departure, who receive attention as the agents of change. Lately, however, there has been rising interest in what was for a time a forgotten era. In 1988 the territorial government opened a historic park at the site of one of the old whaling stations in Cumberland Sound, and other such schemes are under consideration. These are timely developments; for the moment there are still Inuit who remember that "long before the white Canadians came, the Americans and Scots went through their ordeals up here." For the moment there are still Inuit who recall the times "when the whalers and Inuit worked together, helping each other out."

The effect of the years that the whalers spent among the Inuit is perhaps still not fully realized. "We used to call the whalers the Arctic Postmen, because they brought many things," one survivor of the whaling days told me.[4] The whalers brought new equipment, new concepts, new attitudes, new delights. They brought rifles, ammunition, and wooden boats (replacing bows and arrows and skin boats with sails of intestine); they brought metal objects, wooden objects, tea kettles, ships' biscuits, glass beads, and cloth. They brought new music; the accordion rang out from ships afloat or wintering over. Even today's young people sing *O Isaccie!* and the slightly ribald lyrics ("Isaccie's little penis is terribly ticklish"), set to the music of *O Susannah!* They brought wage labour, new diseases, and new genes. The relationship between Inuk and whaler was close and intimate; it is a rare Inuk who has no whaler ancestor. Said an interpreter who helped to collect the interviews on which this book is based, "It would certainly be a surprise to some people down south to know they have so many relatives up here."

Recently, when I sat on a bed writing up my notes in an Inuit household, my hostess asked me what this book would be about. Without waiting for an answer, she suggested, "How about change in the North?" This book is about the start of change. The whalers were the start of change.

FOLLOWING PAGE:
Arvanniaqtut [Whale Hunters]. Simeonie Quppapik, Cape Dorset. Stonecut, 1986.

Introduction   xvii

WHEN THE WHALERS WERE UP NORTH

Prologue

The Arrival of the Whalers

Offshore from Repulse Bay in the upper corner of Hudson Bay lie the Ships Harbour Islands, where wintering whalers left their monument. Chipped into rock beneath the ice and snow lies a giant bowhead whale; nearby are great circles – compass roses encircling memorable names: the *Perseverance*, the *Era*, the *Active*, the *Albert*. Joe Curley, who knew the whalers in his childhood early in the century, recalled, "They'd log their arrival on the rocks ... you'd often see them chiselling away."

In Hudson Bay this was the whalers' "farthest," as early explorers used to say, and early and late in the whaling days the safe harbour for many whaling vessels that travelled through Hudson Strait and passed Marble Island into the great whaling theatre of Roes Welcome Sound. But commercial eastern Arctic whal-

I saw this Whaling Ship. Nicodemus Nowyook, Pangnirtung. Stonecut and stencil, 1976.

ing, the great international industry that spawned wealth for Britain and for the American eastern seaboard, began high on the east coast of Baffin Island, far across the map from Roes Welcome Sound.

The ships that brought the future were the *Elizabeth* and the *Larkins*, out of British ports.[1] They appeared suddenly in 1817, north of Baffin Island in Lancaster Sound. For over a hundred years there had been whaling in the Davis Strait in the waters off the Greenland coast, but previously the whalers had not found a path through the dangerous ice-pack of Baffin Bay. Soon other British whalers followed, daring the dangerous passage to exploit the waters around North Baffin.

Stories of the first contacts have spread all along the old whaling route: "Over there in Pond Inlet," a woman from Repulse Bay relates, "when the first ship was approaching, the people were terrified. They had the shamans then, and the shamans went into trances and chanted. The Inuit thought the people on the ship had come to murder them. When the white peo-

When the Whalers Were Up North

ple came to shore, no one spoke or did anything, all were so terrified.''

The intruders handed out biscuits and tobacco and traded for belongings of the Inuit people. But the Inuit remained afraid. ''That night nobody slept because of the ship. With the daylight the qallunaat – the white men – came back to shore and again they gave out so many things. The Inuit wanted to receive these things but they were still afraid. But the shamans cast a spell on the qallunaat, and they could then do nothing. The Inuit were no longer afraid, and they returned with the qallunaat and boarded the ship.''[2]

For a hundred years there were to be whalers in the eastern Arctic.

Whalers whaled in the Canadian eastern Arctic for a hundred years. The Dundee master John Murray took this photograph of visitors to his vessel the Albert *at the end of the era in 1919 in Albert Harbour, about twelve miles from Pond Inlet, where free traders, who succeeded the whalers and were after furs as much as whale-oil, had set up a trading post. In the background is Mount Herodier, which local Inuit call ''the big stove'' because ''there's always a cloud like smoke over that peak and because they used to boil blubber there.'' In the centre is Qulittalik, the wife of Tam Koonoo, an Inuit whaling mate. ''Before she became his wife she was really poor; afterwards – covered with beads.''*
From John Murray's photograph album, courtesy of Austin Murray.

When the Whalers Were Up North

In Cumberland Sound

In photographs the small station buildings stretch in their isolation across rocky, barren terrain, their wooden structures covered by canvas fastened by wooden battens. "How tiny the houses were," says Leah Nutaraq, a survivor of the whaling days who now lives in Iqaluit, formerly Frobisher Bay. "They look as if there were not enough wood to build them any bigger." But sharp eyes pick out the white man's living quarters, the storage for sealskins, for spare parts, the Mission at Blacklead, the biscuit house at Kekerten.

"Even up till today we call Saturday 'Sivataqbik' – the day of the biscuits," says Nutaraq (her baptismal name is rarely used). "But we received more than biscuits; it was payment day. The qallunaaq would ring a bell to gather all the people together, and if the husbands were away after the whales, the women would go to get their supplies. We got coffee, molasses from a barrel, and biscuits as payment. The

The Ship is Here. Pauloosie Karpik, Pangnirtung. Stencil, 1988.

Kekerten whaling station in its last days, about 1922–23.
From the album of Captain Edmund Mack. Notman Photographic Archives,
McCord Museum of Canadian History, Montreal.

qallunaaq would get out of the house on to the porch and ring his bell, and the women would go up and gather around. When it was time to trade sealskins, he would do the same."

Nutaraq and I spoke in 1987. Her words echoed those of the young anthropologist Franz Boas, who more than one hundred years before described new customs introduced into Eskimo life by the whalers in Cumberland Sound: "Every Saturday the women come into the house of the station, at the blowing of the horn, to receive their bread, coffee, sirup, and the precious tobacco. In return the Eskimo is expected to deliver in the kitchen of the station a piece of every seal he catches."[1]

No one knows for sure how old Nutaraq is, but she was young in the early days of the century and bore her first child during the First World War. She is one of the ever-fewer group of elders who as children experienced life at the old whaling stations that Americans and Scots set up in Cumberland Sound around 1860, the heyday there of bowhead whaling. Around these far-distant outposts of British-American culture, Inuit and whalers built a distinct society, little remembered or talked about in the whalers' homelands but to which *living* memory, as Nutaraq shows, can still attest. Three of us, Ann Meekitjuk Hanson, who is going to interpret, her little daughter, and I have crowded into Nutaraq's small bedroom in hopes of hearing her stories of the last whaling days.

In Nutaraq's youth there were three whaling stations in the Cumberland Sound and adjacent territory: Blacklead Island or Uumanaqjuaq – "like a big sea mammal's heart" – where Nutaraq's family lived ("In the distance the hills are formed like a heart – that's why it's Uumanaqjuaq"); Kekerten ("island") across

the sound, and Cape Haven at the old Inuit camp of Singaijaq at the mouth of the sound facing the Davis Strait. "All the people in Cumberland Sound in those days lived in just those three camps," Nutaraq remembers. At the turn of the century only Cape Haven remained in American hands; at Kekerten and Blacklead the American firm of C.A. Williams, in business in the area for some thirty-five years, had sold out their interests to the Scots competition, Messrs Crawford Noble of Aberdeen, in 1894. "I only remember the Scottish whalers," Nutaraq says. "My qallunaaq father was one of them – I often wonder if I have any relatives in Scotland."

Nutaraq's eyes light up when she recalls the excitement and ritual that attended what was always the greatest event of the year – the arrival at her island home of the ship that brought the biscuits, the Scots freighter that was the supply ship. "In the summertime, when the Inuit were expecting the ship to come in, they would wait and wait on their island. They would wait for a south wind and then they'd walk zigzag over the land, sniffing the wind. If they smelled smoke, they'd say, 'The ship is on the way! It will be here tomorrow.' Then, before the next day came, they'd have a big celebration. Yes, there must have been shamans practising their arts because I was born when there were shamans still in practice.

"At first we'd see only the smoke out on the horizon; then after a long while under the smoke you'd see the ship. I remember how excited the adults would be – because now they would get tobacco, they would

When the Whalers Were Up North

get tea, they would get bannock, all the good things they'd been missing all winter."

The supply vessel arrived about the middle of August and unloaded fuel, provisions, and trade articles, and went on to supply the station at Kekerten, on the other side of the sound. In September the vessel departed, taking with it sealskins, furs, and whatever oil and baleen Inuit whalers had taken in the spring and fall whale hunts. "I myself have never heard of a qallunaaq going after the whale," remarked Etooangat Aksayook, a survivor who lived at Kekerten and who as a young boy went along as an extra hand on one of the region's last whale hunts. "Because the Inuit had learned how; they could manage everything in the operation." The written record supports him. "At the present day, the only profitable and successful manner of carrying on the whale fishery seems to be by the establishment of sedentary stations, managed by an experienced whaler and employing the natives to do all the work," wrote William Wakeham, the officer commanding a Canadian government expedition to Hudson Bay and Cumberland Sound in 1897. "The Esquimaux, or Innuits as they prefer to be called, are first-class boatmen and get to be quite as expert as white men in the use of the modern whaling tools."[2]

When vessels arrived at Blacklead, Nutaraq was often invited to dance for the captains. Tapping her foot and moving her shoulders as she delighted in the memory of the music, she told us, "When ships came, before we went on board, my grandmother used to say, 'If they ask you to dance for them, dance so they'll like you!' My grandmother said this because afterwards we would receive treats – biscuits, bread with molasses. The ships had really nice music. The squeeze box – it sounded so good! I didn't want to disappoint my grandmother so I'd follow the rhythm and dance. I was a very small girl and I wore a sealskin parka and sealskin pants. Afterwards they'd give me bread and butter. On one ship there was a man who would tease and make fun and chase me round the deck. I think he was the stoker of the ship. This man who was so kind to me must have had children of his own.

"I danced for many captains. I was the only child they asked. People had heard that I danced so they'd say, 'Do you want to dance?' There I'd be, moving around and not shy at all. It was a time to be shy, but I hadn't learned that."

By Nutaraq's count, in her childhood thirty-two Inuit families lived at Blacklead. "All those families were working for the whalers," Nutaraq says. "When the whalers stopped coming, they had to scatter to new camps. There were people of all ages – old, middle-aged, young. The very old people were alive when the first white men began to arrive up here." Some of the people Nutaraq remembers met the explorer Charles Frances Hall. Their names run through his famous book, *Life with the Esquimaux*.

The campers lived in qarmait – tent-huts made with wood supports covered, in Nutaraq's youth, by sealskins or sometimes parts of old canvas sails and held

When the Whalers Were Up North

in place along the ground by rocks. In earlier days the beams were whales' ribs. "But in my time we didn't use the whales' bones for support – except sometimes people put two ribs together around the entrance of the qarmak because it was so beautiful there – it fitted so beautifully. But it was more like decoration. It was convenient to use it that way, but we had plenty of wood. Even our grandparents didn't use the whale beams because we got enough wood from the whalers."

Though the passage across Baffin Bay was fraught with difficulty in the wake of the *Elizabeth* and the *Larkins*, British whalers – many of them Scots – came to the North Baffin waters in armadas. Seventy, a hundred vessels some years in the 1820s and 1830s, left British ports. They fished first off Pond Inlet and Bylot Island, then as the season progressed pushed into Lancaster Sound and south down the East Baffin coast, "rock-nosing," as the whalers put it, close to the craggy shore. At first encounters with the Inuit south on the coast were sporadic, but from time to time Inuit and whalers met: "People thought their

Dundee steam whalers in Dundee docks. The British began to equip whalers with auxiliary steam engines that supplemented wind power in 1857, and by the end of the 1860s almost the entire fleet had been converted. Steam facilitated Arctic voyages, increased the safety of vessels in the ice, and extended the duration of eastern Arctic whaling.

Dundee Museums.

tobacco was dried meat; it made them sick. They thought the soap was caribou fat – kind of a Crisco. And a shaman traded with a qallunaaq for a clock. Because he was an angakkuq, he thought the tick-tick was a devil's spirit with a magic more powerful than his. He threw the clock into the water."

But the Inuit soon learned to put high value on the white man's goods. "People started to find out that the items the white man brought were useful. Slowly, not immediately, I think, they learned how to use them," says Cornelius Nutarak, a Pond Inlet resident.

At first the catches off Pond Inlet had often been astonishing – a whaler might catch twenty whales or more in a season. But ice conditions were so severe that whaling voyages there were always hazardous, with many vessels lost. Within twenty years the stocks of whales were reduced and there were fewer whalers sailing annually to the Baffin shores. Then, in 1840, the famous Scottish whaling master William Penny, led by rumour and report and his Inuk mapmaker Eenoolooapik, entered the great broad fjord of Cumberland Sound. Here he discovered abundant riches. A few whalers continued to rock-nose the East Baffin coast throughout the whaling era, but within a short period the sound was the principal eastern Arctic whaling theatre and an icy proving ground for the new techniques and practices that were to usher in a new era in Arctic whaling.

In 1851 Americans who had joined the hunt – Americans were then the pre-eminent whalers of the world – introduced the practice of "wintering." In a bold move a crew from the whale ship *McLellan* wintered ashore in the sound in huts with timber roofs; the men lived off the land on native foods and emerged safe and sound in the spring.[4] Soon both British and Americans began to extend the season by freezing in their vessels beneath the sound's craggy cliffs, sometimes amidst icebergs with polar bears on the summits towering above their masts. In 1857 Captain Penny erected the first land stations, and within a year or two Americans did the same.

Wintering brought about the interdependence of Inuit and the whalers.[5] It altered the way Inuit lived and the way whalers whaled. The Inuit left their small scattered camps where bands of relatives lived together to gather at winter harbours and at the shore stations. They put at the service of the newcomers their knowledge of the land, their seamanship, and their labour in the whaleboats. They provided fresh country food and warm, handmade fur clothing. In return the whalers provided their astonishing firepower and their southern goods. Whaling became a mutual endeavour. In the last two decades of the century, when whale stocks became depleted and few vessels wintered, the stations continued to be the focus of activity and life in the sound, but they remained economically viable because Inuit now did the whaling and harvested blubber skins – sealskins with blubber attached, which were processed in Scotland – ivory, narwhal tusks, and furs.

After Americans in the sound declared the whaling done, Scots at Kekerten and Blacklead continued for

some two decades to find a way to make a modest profit.

William Wakeham, visiting the stations in 1897, found himself impressed. The station at Kekerten, very similar to that at Blacklead, he reported to the Canadian government, had a "well-built dwelling house with capacious store rooms and work shops; half a dozen large and highly finished Scotch built whale boats and a most complete whaling outfit, all in the most perfect order. The boats were up on skids and were painted and varnished; the oars, gaffs, etc., scraped and whitened as perfectly as those of any man-of-war gig; the bomb guns, harpoons, lances, spades, and all the tools pertaining to a whaling outfit, neatly racked, polished and shining. The whole in the most man-of-war order and perfection; all this the work of the natives under the direction of Mr. Milne."[6]

Stations were usually manned by two white residents, the manager, who relied on an Inuk boss to pass on directions to the Inuit campers, and the cooper or barrelmaker. The boss in Nutaraq's childhood was Paula Roach, who for a time had lived in the south with his American father. Some Inuit recall that by their standards Paula "could hardly speak Inuktitut at all." Nutaraq's father was the cooper. She says, "His name was Tom. Inuit called him Coopialla. He was one of those placed there to watch over the whalers' things – the boats and all the belongings."

During the nineteenth century the whalers developed a lethal arsenal to dispatch their prey. In early whaling days the harpoon with rope attached was used to secure the whale, the lance to dispatch it. Typically, after the harpoon struck, the whale sounded, pulling its persecutors on one of the famous "Nantucket sleigh rides," sometimes into the midst of dangerous ice floes, where the whale had its greatest chance of escape. The boat crew poured water over the ropes running over the chocks at the bow to prevent fire. The whale stayed submerged for perhaps half an hour, sometimes longer, but when the giant mammal broke the surface for air, the whalers had their opportunity to make their kill. Other whaleboats from the whaling vessel would arrive to help. But as Daniel Francis writes in *Arctic Chase*, "During this time the boats might be towed for hours, ending up many kilometres from their ship, and often men were lost when they pursued the hunt in a dense fog beyond hearing of a ship's bell and cannon."[7]

By mid-century, however, new weaponry was altering the odds in the whaler's favour. In 1850 a shoulder gun (or bomb or rocket gun) was invented that fired an explosive missile with a time-fuse attached. This was used instead of the lance to deliver the *coup de grâce* to the harpooned animal. The Scots also used a weapon called the harpoon gun ("like a little cannon," Inuit remember), which was mounted on the bow of the whaleboat and fired an armed harpoon. The harpoon gun was not popular with Americans, who considered that it did not work well because of the action of the boat on the water,[8] but in the 1880s they developed the darting gun, a handthrown harpoon with explosive charge.

Station life followed a routine dictated by the seasons. Inuit whaled in the spring and the fall, when the whales migrated in and out of the sound, hunted for seal year round, and went inland after the caribou for their winter clothing supplies in summer, although about six Inuit workers remained at the stations at all times, a survivor of Kekerten station life recalls.[9] Remembering her childhood at Blacklead Island, Nutaraq says, "Starting in May they would begin to get ready, cleaning the boats. They would work for two months. Paula and my father were in charge. When they were whaling, they worked really well. No one treated anyone badly. The bosses were good to the workers, and the workers were good to the bosses."

Each Arctic spring when all was prepared, the Inuit whalers journeyed down to the edge of the ice floe for floe-edge whaling, taking the whaleboats on extra large sleds specially built for the purpose. As Nutaraq has become for Blacklead Island, so Etooangat Aksayook is the principal spokesman for Kekerten. He wrote his own account of floe-edge whaling in an article entitled "Whaling Days" in the October 1987 issue of *Isumasi – Your Thoughts –* published by the Inuit Cultural Institute in Eskimo Point: "Each of the huge boats was loaded onto a qamutik to be taken down to the floe edge, and lots of people would ride inside it. The rest of the gear that was used to hunt whales such as thick ropes, killing spears, harpoons and many other items, was carried separately by dog teams hauling other qamutit. The route to the floe

When the Whalers Were Up North

edge was very smooth because of the flat sea ice. It was a long haul until they reached the floe edge ...

"Once we were on the floe edge, the boats were either pulled up on the ice ready for quick launching or their crews patrolled up and down the coast in the open water in search of bowhead whales ...

"As there were still occasional snowstorms, we would pull our boats up on the ice and sleep in the boats since they had canvas decks over part of them ...

"The crew was instructed to remain with the boat always and to be prepared to take off at a moment's notice in case a whale was sighted near the floe edge. All the crews did the same thing. There were six people to a boat, each one necessary to the others because they all had specific jobs to do. Sometimes they would keep the boat pulled up on the ice, but they were always ready for instant action." After a whale was taken and moored "not easily" at the edge of the ice, Etooangat recalls, the Inuit whalers would "shout Hurray! three times because the whale was dead and ready to cut up."

A constant watch was kept for whales at the stations. "On the island there was a special hill where people spotted the whales," says Nutaraq. "There

Hunters of Blacklead. Among them (left to right, top row) are Qupiruaq, Kakee, Tikiqtituq. About 1922–23.

From the album of Captain Edmund Mack. Notman Photographic Archives, McCord Museum of Canadian History, Montreal.

Blacklead Island whaling station, 5 September 1903. Taken by A.P. Low, officer in charge of the Dominion Government Expedition to Hudson Bay and the Arctic Islands on board the DGS Neptune, *1903–04.*

National Archives of Canada, negative 53579.

was a big barrel there with a little seat – so the whales wouldn't see you, I guess. I'd like to take you up there and show you. There was a place for the telescope and a place for the flag. When a whale was sighted, they'd run up a flag. There were four boats travelling looking for whales, and they'd always be watching for the flag." At Kekerten it is remembered that the lookout was a cripple who used to go to the post, put up a tent, and stay there through the whaling season with telescopes and flags for signalling.

To this day elders recall the taste of the maktaaq, the whale's skin. Eaten raw or boiled, the bowhead maktaaq was a delicacy to Inuit surpassing all others. But it was what lay under the maktaaq that, through most of whaling history, principally interested the whalers. The skin of the bowhead covers a layer of fat or blubber twelve to eighteen inches thick. A bowhead (usually some fifty feet long, occasionally more) may yield thirty tons of oil if a female, twenty tons if a male. And in its awesome head, a third of the animal's size, hangs a ton or more of baleen, slats of keratin, the material of human fingernails, through which the whale strains its food. For centuries whale-oil had supplied light, lubricants for manufacturing purposes (in Dundee jute manufacturers were principal financiers of the whaling fleet), and in the later nineteenth century baleen was the spring steel of the era.

One of Nutaraq's earliest memories is of a huge whale beached on the shore. "There were ropes keeping the whale's mouth open and two people in the mouth. I asked my mother, 'Do whales eat human beings?' 'No,' she said, 'That whale is dead; the people are cutting out the whale's tongue' [the baleen, in the latter days of whaling the whale's most valuable product]. It was the first whale caught that season, and I remember the celebration – how happy they were."

The Anglican missionaries were already in residence on Blacklead Island when Nutaraq was a child. (They had established their first Baffin Island mission there in 1894.) She remembers climbing on stones outside the window of the mission with other little children to watch the Rev. E.J. Peck packing their Christmas presents. Nevertheless, as Nutaraq puts it, the shamans were "still in practice." She says, "I remember being at one of those celebrations when the ship's smoke was sighted. Yes, it is true that there were chantings by the shamans so that there would be more whales for the whalers. The animals were not so easily available to the Inuit, so the shamans had to help. The Inuit wanted to make the captains

The station house at Blacklead Island. In the whaling days women used to gather each Saturday at the station house to receive weekly rations of coffee, molasses, and biscuits. "It was all gone by Sunday," says Leah Nutaraq. Bought in 1924 by the Hudson's Bay Company, Blacklead station was run as an HBC outpost until 1931. About 1922–23.

From the album of Captain Edmund Mack. Notman Photographic Archives, McCord Museum of Canadian History, Montreal.

When the Whalers Were Up North

happy, so the shamans practised their arts to make the whales give themselves to the hunters.'' And reflectively, she adds, ''That's what they did. They were not thinking of evil. One spirit has made the animals for the Inuit to eat. So that's how they practised. They were not committing any crime.''

Commercial whaling in the eastern Arctic is usually considered to have come to an end during the First World War. Nutaraq has good reason to remember how she learned that war had broken out. Her first child was born on her father-in-law's whaleboat out at sea, and she had her newborn son Markoosie up on her back when the whaleboat encountered other voyagers on their way to Cape Haven, the old Inuit camp of Singaijaq, at the tip of the Hall Peninsula facing the Davis Strait. ''There had been a shipwreck and four men were being taken to Singaijaq so they could catch a ship and go to war. Two of the men were crying because they were afraid to go to war."[10]

The Cumberland Sound stations, now in the hands of Dundee interests, were supplied only sporadically during the war, and Nutaraq says, "When the whalers did not come back, we had nothing left. Before the white men came we had relied on bows and arrows. We had used the leg bones of the caribou for arrows because they were so sharp. Even in the first whaling days we relied on these weapons. But now there were no qallunaat. There was no more flour, no bread. Pitaqangi! Nothing! No coffee, no molasses, no biscuits. We used natural teas from the land; they were good, too. There were no bullets, but they searched for tin cans and out of them they got the material for bullets."

There were some whale hunts after the war, but now there were new players in the sound. A number of small concerns, usually called the "free traders," came to the sound hoping to harvest furs, oil, and a variety of Arctic products, and life at the stations continued for a few more years. "We didn't leave right away," says Nutaraq. "We stayed and stayed for a long time. At that time we had Paula as our leader. Then some families took up their own camps. People set up wherever they wanted."[11]

By 1924 the free traders had sold out to the powerful Hudson's Bay Company, which in 1921 had established a fur-trading post in the sound on the Pangnirtung Fjord. Within a few years the Inuit whalers were trappers pursuing white fox and had left their island homes for small camps around the sound.[12]

Not until the 1960s would most Inuit of the sound again leave their small family hunting camps to live together in larger communities – this time with schools, nursing stations, qallunaat housing, and government services – which the presence of the southern technological society in the Canadian Arctic in the years after the Second World War was eventually to bring about.

Nutaraq's reminiscences give us glimpses of the special world that the Inuit and the whalers built in Cumberland Sound. But the sound was only one focus of eastern Arctic whaling. Once wintering techniques had proved their worth there, whaling journeys to Hudson Bay through the ice-clogged waters of Hudson Strait became a practical proposition. The ever-present need for new whale stocks (whalers habitually left fished-out waters behind) had led the whalers from the waters off the east Greenland coast – the East Side, as the whalers put it – across Baffin Bay to the West Side and the eventual discovery of Cumberland Sound. The same need led whalers on round the corner, figuratively speaking, from Cumberland Sound into Hudson Strait and the Bay. The whalers took with them their wintering skills gained in Cumberland Sound, and the expectation of alliances with the Inuit people.

When the Whalers Were Up North

Boiling blubber at Blacklead station. Fires render the blubber in the large metal containers, and the oil is strained through pipes with sieves in the head into large tubs. From these, after cooling, the oil is strained into barrels for shipment. The pipe at the left is probably in the upright position, "just for the camera," says Leah Nutaraq, who watched these operations in her youth.

She notes, "The oil was really clear with not a speck of dirt – just like a Crisco. The bad parts of the fat that couldn't be made into oil were used to make the fire." In the station storage were "stacks and stacks of sealskins – you couldn't count them, there were so many." About 1922–23.

From the album of Captain Edmund Mack, Notman Photographic Archives, McCord Museum of Canadian History, Montreal.

When the Whalers Were Up North

CHAPTER 2

In Hudson Bay and the Strait

"They would paddle to go whaling. The ship would stay put and the boats would approach the arvik – the bowhead whale. There used to be more than one boat going at the floating creature. There were no motors then, so there was nothing to distract the whale; the boat could get right close."

Agee Temela of Lake Harbour is seventy or more and was a baby asleep on her mother's back on the last of the Scottish whaling voyages. She is describing how, in the days when the Inuit used to be whalers, her father and uncles would row up to the great whale in small boats among the ice floes and kill the whale with weapons the foreign whalers brought – the bomb lance and the darting gun.

"I've heard the people from the *Active* would hunt the whales with powered harpoons. My father used to row the boats, and he used to say that as he was

Inuit women and children on deck. George Comer, 1900–05.
Mystic Seaboard Museum, Mystic, Conn.

paddling he would watch the person steering; he could tell by looking at the face of the one who steered the boat when they got near the whale. They would row right close so as not to lose the whale. The person at the head of the boat would harpoon the whale, and once the harpoon hit, you could hear something blasting – inside – to kill the whale.

"There was a very thick rope attached to the harpoon. The whale would pull the boat very fast, and the rope would spin out from the boat. There was something in the bow that kept the rope. They had to keep putting salt water on the side of the boat where the rope went over to keep it from catching fire.

"The boats with my father and relatives never capsized. They never got scared. They just rowed hard at the whale. I used to hear that the whale would make the water crash around it and make big waves."

Agee's story is the first I've heard in Lake Harbour, a community that whalers founded. It was close to Lake Harbour – at the Upper Savage Islands, North

Bay, and especially Big Island – that the whalers met their first Inuit on their journey through Hudson Strait to the Hudson Bay whaling grounds. Earlier in the day the settlement secretary had hefted my bag into a pick-up truck and we had driven down a steep road into the hamlet to the small transit centre that the community maintains for visitors. Now my interpreter and I were seated comfortably with the tape recorder at Agee's kitchen table, drinking a cup of tea. Generations of Agee's family worked for the whalers, both Americans and Scots. Although people along the strait and in the Bay remember that "Americans were first up here with the Inuit," when the elders I met were young, the Scots had become an important presence, making the last days of whaling filled with incident for Inuit of the region. In fact, when in 1903 the government of Canada moved to exercise greater control over its vast Arctic territories – acquired through treaties with Britain in 1870 and 1880 – and established the first police detachment on the west coast of Hudson Bay, it was quickly noted that Inuit families had become known as Americans or Scots, depending on which set of whalers they worked for.[1] (Canadian vessels went to the southern seas but never whaled in Arctic waters, although vessels from Newfoundland, which joined Canada only in 1949, were sighted in the eastern Arctic.)

Pinned to the wall above Agee's head is a photograph of her father taken by Robert Flaherty when he visited her family's camp in the winter of 1913–14, just after the last Scots whaling voyage into the Bay.

Perhaps Agee is one of the babies decked in beads (the whalers' gifts) who appear in Flaherty's photographs but have no names.

Agee grew up listening to her father's and uncles' stories of the foreign whalers; but she heard, too, how Inuit had always caught whales, even before they had fire-power. Some say an ancestor of Agee's late husband was the last man in the area to pursue the whale with all-Inuit technology. His name is known – Lucatsee – and some details of his battle: "It was a windy day – maybe he lost the whale but got some *maktaaq*."

When Hudson Bay whaling began, the classic whaling era was coming to a close. Petroleum would soon negate the need to chase the whale, new industrial products replace baleen. In the future the factory ship in new waters and in pursuit of a new whale, the blue or rorqual, the largest creature known to man, would take from the whale its fighting chance. But in the mid-nineteenth century, when the whalers sailed into Hudson Strait, hunting the whale took skill, courage, and – as some Inuit recall – magic.

The whalers arrived in force in the years after 1860. In that year the *Syren Queen* and the *Northern Light* out of New England ports made pioneer journeys through the strait and into the Bay to look for new whaling grounds. They found them in Roes Welcome Sound, and suddenly there were whole flotillas entering the strait.

Prior to the arrival of the whalers, some Inuit along the strait and in the Bay might have seen a rare explorer's vessel or supply ships of the Hudson's Bay

When the Whalers Were Up North

Company – the Governor and Company of Adventurers of England trading into Hudson Bay, chartered in 1670 – travelling to Churchill and other company posts in the southern part of the Bay. But suddenly these foreign vessels with the tall masts which Inuit thought looked like sleds turned up on end were common sights.

"The first whalers here spoke American," people of Hudson Strait say. The vessels that appeared in the strait in the 1860s sailed up from New England ports, chiefly from New Bedford and New London, and sometimes down from an operation centre in Cumberland Sound. On the Baffin east coast Americans were always an important presence, although British whalers, especially Scots, were more numerous and continuous in their activity. But while from time to time British vessels were sighted, for almost forty years Hudson Bay whaling was essentially an American enterprise. There are records of about 150 voyages into Hudson Bay between 1860 and 1915; of these 59 took place in the first hectic decade, 1860 to 1870.[2] American vessels were sailing vessels. With the Scotch development of steam propulsion in the 1850s auxiliary steam engines rapidly came into use, but American Hudson Bay whalers, practised and skilled in wintering and so prepared to winter over and to have the advantage of two summer seasons in the Bay, remained under sail. Steam whalers appeared in the Bay at the turn of the century when the Scots took up a role there.

First contact between whalers and Inuit occurred a few days' sail (if the weather was good) into the great thoroughfare of Hudson Strait. At the customary rendezvous points near where Lake Harbour is today (and where the Scots early in the twentieth century put up a depot) Inuit from all along the South Baffin coast gathered to wait for the whalers' vessels. The ice pattern and strong natural currents brought the whalers close to shore, and Inuit and whaler would meet, make deals, and fraternize. According to Peter Pitseolak of Cape Dorset, "The people went there to wait because they wanted ammunition and food – whatever the ships could give them. They traded with skins."[3] The vessels sailed on for Hudson Bay, continuing through the Strait, rounding Southampton Island, passing Coats Island, and arriving in the waters off Marble Island below the sound. Then they pressed into Roes Welcome Sound, the channel between the Keewatin mainland and Southampton Island, where the richest catches were taken during Hudson Bay whaling days. Here, whaling vessels tacked back and forth ready to lower whaleboats whenever the whale's spout was seen rising twenty feet in the air, and from the lookout the thrilling cry rang out: "There she blows!"

At the head of the sound, where Repulse Bay is today, were the vast territories of the Aivilik people, mighty hunters, great whalers, who became the principal helpers of the foreign whalers in the Bay. When British whalers reached the East Baffin coast early in the nineteenth century and the first whalers anchored off Pond Inlet, the Inuit there were wary of the qallunaaq and his intentions. But the American whalers who first sailed into the sound and met the Aivilik got a friendly reception.

In many Inuit families there are stories of encounters with the first white man. "Everyone has these stories," Joe Uluksit of Whale Cove told me. "All the stories are just a little bit different, but usually it's the same story with something added." With his wife, Mary Jane Ford, interpreting, Joe told a story that is part of his own family's tradition.

"The first Inuk who saw a white man came home and said, 'Today I saw something with long legs and long arms I never saw before.' He didn't know what it was. It was the first white man. After this Inuit people wanted to know whether the qallunaat were dangerous. Were they scary or not? So an Inuk shaman and a qallunaaq shaman met in an igloo. The Inuk made the white man appear. Both were angakkuit – shamans – and each tried to see behind the other's back. Yes, because they were angakkuit they had something on the back.[4] The white man tried to see what was on the Inuk's back, and the Inuk was dodging around trying to see what was on the other man's back. There was something there – something shiny. My Dad who told this story didn't say what that something was, but in those days most of the people were shamans, so they knew there was something there. The white man threw what he had on his back to the ground, and then the Inuk and the white man went up to each other and spoke each other's language. The Eskimo spoke English, even though he did not know how to speak English, and the white man spoke in Inuktitut. Afterwards neither could speak those languages again. The Eskimo didn't know English, but he spoke it that time talking to that man." When the shamans finished their conference, the Inuk shaman went back to his camp and told his people where he had been and whom he had met. The people believed him because, the story goes, "he smelled like a white man; the whole house smelled of white man."

The great American whaling master George Comer also preserved an account of the moment when some of the Aivilik "first saw ships and white men." He wrote, "It is told of a native whose name was She-u-shor-en-nuck – (he was the brother of the former head of the tribe (Albert) and father of the present chief though the natives do not have a chief they look up to some old man who has always been successful). He had never seen ships or white men, and when he first saw the vessels' masts when two ships were at anchor near Depot Island he could only think and compare them (the masts) to sleds having been stood up but in telling this to the other natives they all went up on a hill and looked and concluded it must be the white man's ships which they had heard of.

"The next morning they started very early and went to the shore of the Harbor where the ships were then. Leaving the others to build snow houses She-u-shor-en-nuck went alone to the vessels where he was well received and given a number of small articles. He returned at night to his people highly pleased with

George Comer in the rigging. Probably taken in 1907 aboard the schooner A.T. Gifford, *which Comer captained after the* Era *was wrecked off Newfoundland on the journey out to Hudson Bay in 1906.*

having met and seen and talked to the white people.

"The two ships were from New London and the captains were Brothers Chappel and Henry Y Chappel ... It was on board of Captain Henry Y Chappel's ship that the natives first saw some pigs or hogs and from these concluded that they were Tonwarks and were kept by the white men to ward off sickness or trouble ... the white men's guardian spirits."[5]

The Aivilik and members of other tribes soon put their skills at the whalers' disposal, in return for weapons, ammunition, and material goods, whaling with them in summer, camping near the vessels in winter at whatever harbour the ships' masters chose. Historians note that the whalers drew even inland Inuit from their traditional lands to the shores of Roes Welcome Sound and to the various anchorages, usually among islands, that became popular. The names Marble Island, Depot Island, Cape Fullerton, Repulse Bay resound through the whalers' journals.

Over the years whalers set up a number of stations and depots. In the mid-1870s American interests explored possibilities in Hudson Strait, and a station established there late in the decade operated for a number of years. The Dundee whalers who became

When the Whalers Were Up North

active in the area just before the turn of the century set up a station on Southampton Island; after it closed they maintained a vessel as a station in the Repulse Bay area. They also had a depot at Lake Harbour. These operations had important consequences for the Inuit, but none achieved the permanence of the Cumberland Sound stations. Throughout Hudson Bay whaling (the last voyage out of the Bay was in 1915) it was vessels wintering over that were the important southern presence and agents of change.

In his journal the explorer Charles Francis Hall described life for the Hudson Bay whalers wintering over in 1865. "In this harbor, a little more than a mile west of Depot Island, four whalers besides the *Monticello* were anchored within rifle-shot of each other; these were the *George and Mary* of New London; the *Black Eagle* and the *Antelope* of New Bedford; and the *Concordia* of Fairhaven, Mass. Each was banked up with snow six or eight feet thick and nearly up to the gunwale, the upper deck being well housed. On board the *Monticello*, although but little coal was used, the temperature was kept above 32° throughout the vessel. Five other whalers, including the *Ansell Gibbs* and the tender *Helen F.*, were anchored in a commodious harbor completely land-locked on the northwest side of Marble Island, an islet about 15 miles in length, lying 12 miles off the coast.

Comer's mate Harry, centre, and other Inuit, 10 March 1905. Attributed to both George Comer and to Frank Douglas MacKean. National Archives of Canada, negative C1516

Men gathered around beached whale. George Comer, 1900–05.
Mystic Seaboard Museum, Mystic, Conn.

"On board all of these vessels the amusements usually gotten up by Arctic voyagers for maintaining the cheerfulness and health of their crews were at this time in full play, and were generally of a theatrical character, varied by masked balls and by several forms of the dance."[6]

All along the whaling route new societies emerged. Inuit received new names. Wrapping tongues around Inuit names was a problem for the foreign whalers, so they resorted to their own system of nomenclature, handing out common English names (Charley, Mary, Harry), or names that identified helpers with specific vessels or captains, or strange nicknames: Starboard Eye, Bye and Bye, John Bull, Jumbo, Stonewall, Shoofly. "In the days of the sailing ships, the whalers started giving out these odd names and since that time those names have stuck." Apparently perplexed by the same problem, the Inuit too went in for name calling. "The whalers weren't too fussy about names in those days, so by what they were doing, so they were called."[7] Physical characteristics also inspired an Inuktitut name for a foreigner. The well-known captain John Murray was nicknamed Nakungajuq – Cross Eyes. ("He did have a cast to his eye," says his son Austin Murray of Wormit, Scotland.)

And a new hierarchy developed. Captains gave their orders through an Inuit mate. In those days many Inuit men had two or even three wives. "Oh, it was necessary for a great hunter; he needed two wives to work on his skins." One of the chief's wives sometimes travelled with the captain. "The captain's girlfriend wasn't always the chief's wife, but she'd be a sister or a close relation."[8]

As they had in Cumberland Sound, Americans and Inuit conversed in pidgin, with words often seemingly from no known language but apparently easy for both Inuit and qallunaat to understand. Joe Ebierbing, the great Inuit interpreter who travelled with explorer Charles Francis Hall, said Inuit arriving from the interior spoke "old fashion."[9]

The whalers brought time, or, more exactly, new gauges of time. They divided their days by lunchtime and dinnertime; they brought clocks and watches.

And they introduced a revolutionary new concept: the seven-day week. "My parents had their own names for months but they had never known days – Monday, Tuesday, Wednesday," the old Lake Harbour hunter Kowjakuluk told me. Scholars have described months and years as natural phenomena, but they believe the week was devised by man for his convenience.[10] Weeks often pivoted around market day. When days of the week became important, Inuit began to fashion their own calendars: slabs of wood with two rows of seven holes. "When they started knowing the days of the week, they used to put a peg in the first hole for Sunday. When they woke up in the morning, they'd put the peg in the next hole."

And in the games and entertainment the qallunaat arranged to amuse themselves through the long winter, the Inuit participated. Aivilik recall that "the square dances started then and have continued to this time ... Sometimes for dances the Inuit made dresses with little puff sleeves."

There were greater numbers of Inuit in the whaleboats in the later years of whaling, but even on the first voyages into the Bay the reputation of Inuit as hunters and seamen was such that they were taken on to fill gaps in the qallunaat whalers' boat crews.[11] "They say some of those Inuit whalers were really good," a whaler's descendant remarked. "In the whaleboats they used to have an Inuit captain. The tradition was, you'd take the man who had the most experience and let him be the leader."[12]

There was some mystification, never quite dispelled, over why the qallunaaq wanted the blubber of the great whales. "Today people ask this question: What did they use it for? Why did they want it? A lot of people have asked. But even in those days they wondered. Because we knew that the whale blubber would not burn fully in an Inuit qulliq – the stone seal-oil lamp – like other fat. Like seal oil. It was not the right fuel for the qulliq. So even to this day people are asking the question: Why did the whalers want the oil and baleen?"[13]

Throughout the long history of European whaling, whale blubber had been a principal source of light. In the nineteenth century, with the discovery of coal, gas, and petroleum, whale-oil as a fuel was supplanted, but it still had industrial uses – as a lubricant, in tanning, in the manufacture of paints and varnishes. But as the century progressed, whalers hunted their quarry less for the oil than for the baleen, the strange strips of bone that hung like a curtain from the upper jaw in the whale's great mouth. Baleen was flexible and easily twisted when heated. It had many uses, including springs, carriage wheels, whips, corset stays, and fishing rods. "The baleen whale used to be very valuable – its oil was, but the baleen was more so," Peter Pitseolak of Cape Dorset observed. "The baleen was huge. If a man carried it on his shoulder, it would touch the ground on either side. It was very flexible. The Eskimos used it a lot for bows and arrows and harpoons. I don't know what the white men wanted it for, but the qallunaat wanted it just as much as the Inuit. It was very useful."[14]

All products of the whale were valuable to the Inuit. They made their sleds from its jaw-bones (the early Eskimo people who didn't use the snowhouse had roofed their stone homes with its ribs). Its carcass was a giant bait that drew polar bears and foxes. Its skin was the most prized of all foods, delicious and nutritious; its meat fed the dogs all winter. A whale was subsistence for a year.

Descendants of the Inuit whalers recall hearing how, when the Inuit first met the whalers, "they were really amazed" at the whalers's equipment. "They were fascinated and astonished. The whalers had those huge weapons – like cannons with harpoons at the end – a huge spear made out of steel."[15] Today even Inuit sometimes wonder how it was possible to kill the whale without the white man's whaling gear. "I can't believe the Inuit hunted whales before the whalers came," says a woman who in her childhood watched her relatives catch whales. "The whale is such a huge animal."[16] But all along the whaling route there are Inuit who still know the details of their old technology.[17]

According to Rosa Kanayuk of Repulse Bay ("I'm never lazy to talk about the whale!"), whether the qallunaat were in the vicinity or not there was never a time when the Aivilik of Hudson Bay did not catch whales. In the folklore of her region even the women caught whales. "There's a place nearby called Situharvik," she says, "and this is how Siturharvik got its name.

"There once were two men who had a kayak but were scared to ride on it. So their two women prac-

When the Whalers Were Up North

tised at night time, staying up late, until they had mastered the kayak. There were lots of whales then in the waters around Repulse Bay, and when people went whale hunting, they'd find the whales just floating on top, taking it easy. The two women started going close to the whales, getting them used to the kayak. Everyone knew the women were going to go whale hunting, so they had harpoons, floats, everything they'd need to catch whales. The whale they intended to kill the women just left alone until it got used to the kayak. The harpoons then were made from tusk and antler; there was no metal then. The women moved closer and closer, and the whale got used to the kayak.

"When they decided they felt brave enough, they threw the harpoon along with the float. Then they chased the whale to kill it.

"Those two women killed the whale alone. The people of the camp were so excited and happy that after the arvik was all cut up, they took pieces of the skin with the blubber still attached and started sliding up and down a little hill. That's how Situharvik got its name: it means a place where you can slide."

Joe Curley of Eskimo Point related, "Before the whalers came, the Inuit people already knew how to hunt bowhead whales. I never knew Seegak; he was

Walrus hunting by sealskin boat. These large boats were both sewn and rowed by women. Pitseolak Ashoona, Cape Dorset. Felt-pen drawing, 1970.

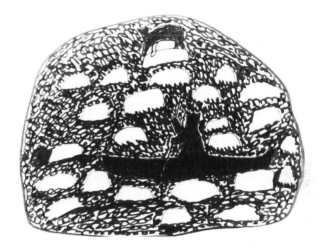

Kayaker among the ice floes. Pitseolak Ashoona, Cape Dorset. Felt-pen drawing, about 1975.

my grandfather's companion, but he was a very very successful hunter who caught whales in the old way, probably before the whalers came. I think he got his other name, Tapatai, from the white people. I am not sure, but I do suspect this. Tapatai is not an Inuit name. He used to go whale hunting just in a small kayak. If you harpoon bowheads, they think they are attacked by sharks and automatically they go towards the shore. There in the old days, the hunters would harpoon the creature constantly with weapons made from tusks. They'd use a narwhal tusk with a sharpened flint stone and it would make an incredible wound on the creature. They had no modern weapons then, so that is the way they had to catch whales."

The hunter Ikidluak of Lake Harbour says, "Before the whaling ships brought their equipment, the Inuit used to hunt from their kayaks. They'd tie a knife to a paddle and stab and stab the animal. Not many, only a few from round here used to do it. Lucatsee killed a whale that way, but there was so much wind that when they tried to tow it towards the land, they couldn't. Maybe they got some maktaaq. Lucatsee was from this area; I saw him when I was a child. His grandchildren are scattered now, but Lucatsee's son had a son who is the father of Soo who lives here in Lake Harbour. She is Agee's daughter."

To catch whales the kayaker rode right up on the whale. "That's when they harpooned. The whale wouldn't go down right away. The whale doesn't move suddenly; there's time to move back. The body of the whale is not dangerous; it is the tail that's the most frightening."

Osuitok Ipeelee of Cape Dorset recalled the unique specifics of the Inuit whaling gear – the harpoon head and the whaling avataq, the inflated float thrown with the harpoon to keep the hunter's quarry on the water. "When I was a boy and falling asleep at night in the igloo, I used to hear stories about the Inuit catching the whale by kayak before the white man was around.

"When any wild animal is harpooned, it may become furious. In the days when the Tuniit hunted the bowhead, I've heard they used to talk to them. They'd talk to them, and they'd calm right down. The Tuniit would say 'Saima! Stay calm,' and it seems the crea-

tures understood. Once a hunter harpooned a whale, the other kayakers would start harpooning, too. They would kill the whale with the harpoons, striking over and over again. The blood would flow inside the whale and the whale would die. The harpoon head had a weak attachment. It would loosen and travel inside the whale. When the whale was in pain and writhing, the harpoon head would travel forward as the muscle moved. It couldn't slip back – it had serrated edges. Each time the whale moved, the harpoon head went deeper. With this kind of head, it didn't take long for the whale to die.

"I've seen this head but made from metal; long ago the head was rock. They'd use the seal's last rib – no other rib would do – to sharpen the edge. Some parts would be shifted away – it was just like using an electric machine.

"The yearling seal was used for the sealskin float. The yearling seal is strong for hunting any kind of animal – walrus, whales. This kind of float is called sagalajaq – a yearling. The bowhead is a huge creature and you might think it would need a bigger float. But maybe it's the same with animals as humans: when we are in our teenage years we are strong and spirited. Perhaps nature provided a magical thing. Sagalajaq is the right float for the whale.

"Even the whale cannot win against its strength."

Oola Kipanik at Lake Harbour had further information. "Yes, they used to hunt whales before the white man came, with kayaks and harpoons and maybe with boats, too, because I have heard my mother tell how when she was a baby upon the back of her mother, they were making boats out of ukjuk skin. My grandmother told her how one of the men carried her on his back while my grandmother sewed the skins for an umiaq – the sealskin boat.

Kiawak Ashoona, another Cape Dorset hunter, explained how whale hunters used the sealskin boat as a drag. ''If they hunted from boats, they usually had a long rope attached to the boat. If they didn't attach a rope, they'd use a piece of board with the avataq made from the baby seal and a rock attached. There'd be a rope to the board. One man alone could get a whale if he used the rock and board. The whale would take himself towards the shore because the rock weighed so much the whale probably feared the rock would sink him. He'd swim in the shallow water near the shore.

''And when they harpooned the whale they were not to shout for joy or show excitement. Even when the whale was bleeding badly, there was to be no joy or shouting until the whale was dead.

''They believed if they got excited, most likely they would lose the whale.''

When the foreign whalers arrived, Inuit weaponry was instantly surpassed. But throughout the whaling years there were other aids the Inuit believed powerful indeed – the ancient, mysterious arts of the shaman.

When the Whalers Were Up North

Shamans and Whalers

"In those days, around the time of Christmas Day, they used to build a giant igloo so they could have celebrations with the shaman people. First, they would hunt together and try to catch food to last for the year. Then five men, perhaps, would build a big igloo for the festival. They would have four quliit [stone seal-oil lamps] around the side and one in the middle. In the giant igloo they would do the singing and dancing, hoping that when they went hunting the singing and dancing would help them draw the animals.

"Yes, it is true we used to have drumming here. For some of the drummings, not all of the drummings, the drummers wore masks from skins with tattoos around the holes. It is true that the drum beat carries a meaning. Each hymn has a different tune, and just like that, the drummings are different, too. Every-thing has a name – cups, lighters, chairs – and just like that, each different drumming has its own mean-ing. The drummer would play for a subject and the people would know the meaning. In the old days when they caught big whales by kayak, they'd drum to draw the animal – to make it easy catching. The same as in the army, when the commander gave the order, there'd be a quick action."

The speaker is Pauta Saila of Cape Dorset. He is describing the world the whalers met when they first sailed through Hudson Strait. Using modern idiom he is telling us of the festival of siiliitut, the great celebration of fecundity and the renewal of life that once took place all over the Arctic at the time of the midwinter moon. For part of his life Pauta's father, like many important leaders, was a shaman, and Pauta is said to have a special knowledge of their powers.

Television flickers in the background as we talk. Through a window you can see the hill where, Cape Dorset people remember, sometime in the sixties the last igloos stood. Pitaloosie, Pauta's wife, watched

This coloured-pencil drawing by Jessie Oonark of Baker Lake appears to show shamans invoking their spirits to ensure good hunting. Collection of Mr and Mrs Andrew Davis.

their lights shine out at night until the occupants, like the other campers, moved into government-supplied houses. I met Pitaloosie, who learnt her English in the south, on my first trip to Cape Dorset, and on more recent visits I've been lucky enough to learn something of the old beliefs, sometimes with Pitaloosie as interpreter, from Pauta's stories of the shamans and the tuungait, their spirit helpers.

Today the Inuit are Christians – Anglicans, Catholics, or sometimes members of new popular fundamentalist congregations, often with Inuit leaders. But when the whalers came to the North, and for long after, shamanism was the heart and soul of the Inuit hunting culture. The shamans were intermediaries between man and the spirit world. They used their special powers to make hunting good, to make animals "easy catching." They gained their powers sometimes by inheritance, through a "calling," by apprenticeship. The shaman had powers to fly through the air, to see faraway places, to cure the sick. There were good and bad shamans, but the good shamans "were Gods to the people."[1]

When the shamans regulated the Inuit world, there were ancient rituals, symbolic responses that speak of the interrelatedness of human and animal life. "If my wife were in labour with her first child," an old hunter explained, "I would loosen my kamik laces. The laces were thin and came from the skin of sea mammals. I would do this so that the first child would come out smoothly and all future children be easily born.

"Years ago, and it probably still happens today, the woman who helped a mother when her son was born would be to that boy an angusiaq. When the child was born she'd say, "When he grows up, he's going to be a very good hunter. He'll be able to catch seal or rabbit and other animals easily.' So when the boy grows up and hunts and catches his first animal, he takes his catch to this woman who was present at his birth.

"Long ago the women used to wear sealskin panties. Panties have a waistband, so the sealskin panties had laces around the waist. When the young man came with his first animal, the woman would pull up the front of her amautiq and she'd say, 'Loosen my laces.' Sometimes the young man would be quite shy. But he would be happy he had caught his first animal, and right away he would loosen the tie around the lady's waist. When the young man did this, it meant that now, when he went hunting, he would easily catch animals."[2]

The Inuit took out the laces when a whale was hunted. "Perhaps this happened, perhaps not," says Pauta, "but this is what I heard from my grandmother. When a hunter on a kayak speared a whale, other kayakers would be nearby. They'd begin to chant – so the whale would not try to get away from the hunter. People would chant in order that the whale would stay in one place – to make it easier to kill – and while they chanted, they took out the laces from their kamiit. They would take out the laces so the muscles of the whale would be loose.

When the Whalers Were Up North

"The people watching on the land would do the same thing. Whenever there were hunters trying to kill a whale, the people on the shore would get together. I have told how in the short days near Christmas the Inuit used to build a giant igloo for celebrations with the shamans. They used to do the drum dance and sing the ai, ai, ai, but they could do that outside, too. Perhaps when they were catching a whale they would do something almost like a drum dance. To get the whales to come to them.

"When hunters killed whales, they would try to spear the whale so it would die quickly. But sometimes the whale would try to get away and swim under the water, pulling the hunters on the kayak.

"When the whale goes under the water carrying the float, that's when people would take out their laces. From the top of the kamik. And if they had them around the ankle, they would take those out, too. That is the way they would wait for the whale to die.

"I don't know exactly what kind of chanting they did. I have often asked myself what sort of chanting they used to make."

Scholars have come to believe that all across polar America similar strict ritual once governed the hunting of whales. "The importance of ritual to Eskimo whalers has long been recognized," wrote J. Garth Taylor of the Canadian Museum of Civilization in 1985. Quoting the Fifth Thule Expedition member, he said, "It is half a century since Birket-Smith (1936:89) observed that among Eskimos, 'There is on the whole no animal [except whales] ... whose hunting is so hedged by strict taboo, magic formulas, and the use of amulets'."[3] After the foreign whalers came, the great hunters who were the Inuit whaling mates were regularly shamans. The famous Inuit bosses Harry and Angutimmarik, also known as Scotch Tom, who whaled respectively for the American captain George Comer and Scottish master John Murray, were both shamans. Scotch Tom, who died in the late 1940s on Southampton Island, is remembered as a great shaman who "may have helped out quite a bit in bringing the whales. In those days they used to practise mind concentration."[4] The shamans knew where the whales *were*. "I've heard a little about the shamans being involved with the whaling," says Osuitok Ipeelee of Cape Dorset. "A shaman wouldn't tell people about his powers, but on a hunt, if a shaman wanted to go in a certain direction, that is the way the people would go. The shaman would say, 'That's where the whales are.'"

Among the shaman's skills was probably the power to hypnotize. People all along the whaling route know this story that Osuitok Ipeelee tells of the great Baffin Island shaman who kept the peace on the Scottish whaler *Active*: "When the people were hunting the whales, they would be gone on the ship for a long time. Years ago on one of the ships there was a white man who was very unpleasant. Just as today, the Inuit ate their meat raw, and when the Inuit were eating the seal meat, he'd go around and criticize and act as if the sight made him sick. He'd pretend to get nau-

When the Whalers Were Up North

seated. When we eat the raw meat we get blood on our hands. So one day when this unpleasant qallunaaq was criticizing, an Inuk made a fist and pushed his fist in the gallunaaq's mouth. He gave him a bloody face. The man went away to wash his face, and after this for a while he didn't come back. But soon he was at it again, criticizing the Inuit as they ate.

"Among the Inuit whalers there was a shaman. The shaman did not speak to this man, but he got control over him, and suddenly this unpleasant qallunaaq started dancing. He couldn't help himself; he was dancing and enjoying himself. The shaman had him under his control and he couldn't stop dancing.

"There were other white men around. They didn't eat raw meat either, but they started clapping their hands because they knew this man was mean.

"Because he was mean, it seems he was always unhappy, and because he was unhappy he'd criticize and criticize. The shaman made him dance, and while he danced, he was smiling and enjoying himself. The shaman made him dance to get him to realize he'd been mean, to make him realize he should enjoy himself. The shaman didn't want to kill him."

The shamans on the whalers are said to have been

Elephant, by Repulse Bay carver, artist and date unknown.
The Eskimo Museum, Churchill, Man.

well rewarded. Maria Teresa Krako in Chesterfield Inlet had shamans in the family. Both her parents were shamans, her father Qimuksiraaq one of the most renowned on the coast. She says, "Probably my father worked for the Americans," and believes that when she was a child her family travelled with the American whaling captain George Comer. She relates, "I have heard of the white man who had magic powers himself, and I have also known many other well-known shamans. I myself was called after the shaman Paalajaaq. I have heard that the qallunaaq used to pay him quite a bit when they wanted the whales to appear. Most of the time when a shaman used to say where the whales were, he'd be amazingly accurate. Because he was a shaman he could do these things."

The shaman relied for information on communication with his tuungaq, the helping spirit who put

Saila and Alainga demonstrate for the camera of American explorer Donald MacMillan how the shamans performed to make animals "easy catching." Photographed at Cape Dorset, 7 August 1922. Peary-MacMillan Arctic Museum and Arctic Studies Center, Bowdoin College, Brunswick, Maine.

Shamans and Whalers 39

him in contact with the spirit world. They say the shamans could make their tuungait out of "any kind of animal, from worms, bugs, and the spirits of dead people,"[5] and the coming of the whalers introduced some new possibilities. The first American whalers who came to the Bay sometimes brought a few pigs or hogs as supplies of fresh food. Inuit concluded they were tuungait, Comer related, "kept by the white men to ward off sickness or trouble ... in other words, were the white men's guardian spirits."[6]

Maria Teresa Krako says her father had "qallunaat animals" – white men's animals – among his spirits. "My father Qimuksiraaq was a shaman, and I've heard he could fly through the air. I know my mother used to ride those big birds, the Canada geese. My father used other animals. He used to say that besides the spirits of our own animals he also used the spirits of animals that are foreign to us, the white man's animals, as his helpers.

"Did the whalers introduce the foreign animals? Possibly, maybe. In fact, when you look at the movies you see that the qallunaat animals are very different. Here we have only a few animals – polar bears, walrus, muskox, and wolves. The wolverines and foxes. And siksiks [Arctic ground squirrels] and whatnots. We don't have that many animals; down south they have all kinds."

Leah Arnaujaq of Repulse Bay recalled that her father Tukturjuk, or John Bull, had a particularly impressive qallunaaq animal among his helping spirits.

"In those days they had no doctors or nurses to cure the sick, so the shamans used their magical powers to help those people who were ill. My father had an elephant as his spirit. Probably he heard of the elephant from the whalers. Cross Eyes – John Murray – used to tell all kinds of stories about the foreign animals ... Probably he had seen an image of an elephant. I really don't know how he happened to have a spirit like that, but he was able to use his magic powers with this elephant. He had all kinds of spirits – the elephant was not the only one – but he used to talk to his elephant spirit often."

Leah Arnaujaq recalls that living with the spirits was not easy: "He used to tell me what he saw, but I have mostly forgotten and I don't want to tell lies. Sometimes I would be afraid when my father was telling his stories, especially since to me the spirits didn't exist. It made my hair stand on end to see my father using his magical powers, talking to spirits all around him. I was terrified, knowing my father was a shaman, and afraid the spirits would come alive in front of me. As soon as possible I would try to forget, but he talked to his spirits quite often."

There is evidence that word of the elephant came to Hudson Bay with the first whalers or very shortly after. In winter harbour at Marble Island in 1864 the crew of the *Orray Taft* provided elaborate entertainment in a theatre they constructed on the Arctic rocks. A success of the season was the "Artic Elephant." His image can still be seen.[7] It appears, sketched in a notebook by an unknown crew member, possessed of ferocious tusks but human feet and draped in qallunaaq material. Across the bottom of the drawing the whaleman has written, "THE ARTIC ELEPHANT off Marble Island. Captured on the flaw by John Bull."

May we presume that soon after the whalers came to the Bay, a shaman summoned the whale for the foreign whalers with the aid of "qallunaat animals"?

When the Whalers Were Up North

CHAPTER 4

The Case of the Missing Whales

I. THE THEFT

"It must be hard to steal whales," said Annie Manning, a modern Inuit justice of the peace, hearing the story of Johnnibo for the first time. Bowhead whales run about fifty feet in length, sometimes more, and weigh in the neighbourhood of twenty to thirty tons. Despite these mighty measurements, in 1880 John O. Spicer of New London believed he was robbed of three whales. Whale-oil that year was worth about half a dollar a gallon, and whalebone or baleen – used for buggy whips and corset stays – two dollars a pound. As a result, distinguished New England attorneys argued the merits of his case in court, and Johnnibo made the long journey from Hudson Strait to a Boston courtroom as witness for the prosecution.

Johnnibo, his wife Annie Kimilu, and her little

New Bedford wharves, about 1870.
The Whaling Museum, New Bedford, Mass.

daughter Kudlarjuk were not the first Inuit to visit the New England whaling ports. They lived in close proximity to Joe Ebierbing and Hannah Tookoolito, who travelled south with the American explorer Charles Francis Hall and went on lecture tours. Annie Kimilu's own father Cudlago died tragically on a voyage back from the United States, asking urgently as he weakened, "Do you see ice? Do you see ice?"[1] By the time of their visit Inuit in both the Cumberland Sound and the newer whaling grounds of Hudson Strait and the Bay were increasingly dependent on the whalers' gear, ammunition, and boats and more and more appreciative of their tea-kettles, beads, tobacco, and music. But Johnnibo was almost certainly the first Inuk from his country to find himself dependent on Western civilization's most vital concepts: common law and due process. And to bear the law's delay.

"Yes, it must have been very difficult for them," says James Akavak of Lake Harbour, one of the early special constables for the Royal Canadian Mounted Police, who in his youth knew both Annie Kimilu and

Kudlarjuk. "The court hearing must have gone on for a long time – I would think the case went on for one or two years. Kudlarjuk was just a little girl when she and her parents were taken south. She had to learn Inuktitut all over again."

Undoubtedly the activities of the whaling days spawned many high dramas now faded irrevocably beyond memory, but stories about this law case that put a Baffin Island Inuk on the witness stand, the events that preceded it, and its aftermath still circulate on Baffin Island. Many details are preserved too in the logs, court records, and newspaper clippings held by American whaling museums.[2]

The Case of the Missing Whales, as interpreters dubbed the events, properly begins in 1877, when John O. Spicer, exploring along the north Hudson Strait coast, met hunters who told him of a place called Akuliak, where, they said, there were always whales.[3]

Captain Spicer ran rather large-scale eastern Arctic whaling operations for the New London firm of C.A. Williams and Co. He had built his own station in the Cumberland Gulf, probably at Blacklead Island, known to the Inuit as Uumanagjuaq – "like a big heart." But he ran satellite stations too, and kept at least three vessels (he owned shares) whaling along the coasts, into Hudson Strait and the Bay, and ferrying oil and whalebone from his stations to his New London, Connecticut, port.

Spicer wintered over near Akuliak, whaled successfully in the area in the spring, and picked the point for a new Hudson Strait shore station. Then,

on his next whaling voyage in 1879, he sailed up to his east-coast stations and around Cape Haven, or Singaijaq, as Inuit say (his station just north of the headland of Frobisher Bay), and sought out Johnnibo, an old acquaintance and probably a regular sailing companion.

To the qallunaaq Johnnibo was "John Bull" or "the Mate." Like other Inuit whaling bosses who had put their skills at the disposal of the foreign whalers, he had his own whaleboats (earned by success in the hunt) and crews. Inuit say his name is a contraction of John Boat. "A qallunaaq was once in the area where John had his boat and said, 'Is that John's boat?' That's how the name came about. It was accidental." He had lived and worked for many years on the Frobisher headland where whalers whaled in the Davis Strait. The explorer Charles Francis Hall met him there in 1860.[4] Like many other Inuit whaling bosses, he may well have been a shaman, a religious as well as a secular leader.

"Engaged John Bull and his crew to go to Hudson Strait to whale for *Era*," reads the vessel's log for 18 July 1879.[5] The arrangement was a rather usual one, known today to whaling historians as contract whaling. "In those days," explained one Inuit informant,

Johnnibo, Annie Kimilu, and Kudlarjuk in New London in 1881. An item in the Mystic Press *for 22 December 1881 notes that Kudlarjuk is "as bright a child as one need to see."*
John Bishop, New London County Historical Society, New London, Conn.

When the Whalers Were Up North

"the ships used to let the Inuit off at a place and the people would stay there through the winter until early in the spring when the ship came back."

All set sail on the *Era* for the new promising waters some five hundred miles away. On board would have been the Inuit wives, the children, the qamutiit – the sleds – and all the dogs. After summer whaling, Spicer dropped Johnnibo and his men off at Akuliak, now a link in his chain of stations. He was outfitted and equipped, it was later said in court, "for the purposes of carrying on the whale fishery … and especially to capture whales … and to keep, take care, and have custody of such whales."

For the Inuit of the coast the arrival of the newcomers must have been an event of great social and economic importance – and was recognized as such. Pauta Saila of Cape Dorset heard the stories that his grandparents Pauta and Peuliak and his father Saila, a child at the time, told all their lives of the first whaling voyage they made with Spicer and Johnnibo. They were local Inuit, picked up on one of the whaling cruises (in either 1877–78 or 1879) that Spicer made at the time he established his station. They became closely associated with the station for a number of years.[6] They lived in vast territories up on the west side of the Foxe Peninsula, but they were summering down on the Hudson Strait, camping with Peuliak's parents, when Spicer and Johnnibo sailed by and picked them up. Also on board was Pitseolakpuk – "the bigger Pitseolak" – who as a child had been placed on a whaler to learn English. He became "the first

interpreter"[7] and one of the first Inuit pilots. "He was a leader around Lake Harbour, and he was along because the qallunaat wanted someone to lead them," says Pauta.

At the time it was the custom to build inuksuks – stone cairns – to attract the whalers' attention. "The Inuit made inuksuks at Tikkuut, a small island near Qimituuq, and there were even more inuksuks at Saunirjuak close by. The reason they made them was so the whalers would know there were people there.

"The ships would pass by on their way to other places and the qallunaat would look for people to help. There would be people staying at Tikkuut and also around Sagva, and every summer people went to Qimituuq because it had so many flipper seals. The whalers used to use Inuit helpers, so when the Inuit wanted, if they didn't refuse, they would go on the ships and work.

"One summer my grandparents, Pauta and Peuliak, with my father Saila who was just a little boy, were asked to come along so Pauta could help the whalers spot bowhead whales. Johnnibo and his family – Annie Kimilu and her daughter Kudlarjuk – were on the ship.

"The ship stopped at Qimituuq and then stopped at Itilliajjuk to ask the Inuit there if there were whales in that area. Then they went up on the west side of the Foxe Peninsula, and they were not far from my grandfather's camp there when a white person died. He died because he could not pee. He had brown hair. They brought him to the shore and dropped him off and buried him on top of the ground in material like canvas. They covered him with rocks. I have seen the place where the white man is buried, but you won't find the body or the bones. The polar bear got in there and tore the canvas and perhaps ate the body. But you can see the circle of stones.

"After they buried the white man, the whalers turned back and went on towards Coral Harbour, passing by Nottingham Island. In that area there used to be lots of whales and the qallunaat did a lot of whaling. Altogether they caught two male whales, which were quite big, two smaller female whales, and a baby whale. After they caught the two big male whales, with the blubber they'd got, the big barrels were almost full. The whalers went home almost right away.

"My grandparents Pauta and Peuliak passed the winter down on the south coast. In those days the ships used to let the Inuit off at a place and the people would stay there through the winter until early spring, when the ship came back. Long ago the ships would arrive when there was still some ice floating. In July. Those ships would have sails.

"I think my grandparents were often in the same place with Johnnibo because Kudlarjuk and Saila, my father, grew up together even though the one was a girl."

The theft of the whales that led to the law case occurred in the spring of 1880. Johnnibo and two boat crews were deployed in the Hudson Strait, whaling as arranged for Captain Spicer. They were successful in the hunt. But the captain arrived late, delayed by bad ice conditions. By the time the *Era* sailed in, the products of the catch of three whales had disappeared, carried off by rival American whalers. "Johnnibo was tricked," an old hunter remarked: "he was told the rightful owner was never coming back; he was told this by people who wanted the blubber."[8]

Spicer discovered his loss on 25 August 1880, shortly after he sailed into his harbour. "Went ashore to look after our station, and found some skipper had been here and taken bone and oil or blubber," wrote a scribe in the log of the *Era*, lying at anchor in Spicer's Harbour, about fifty miles west from where Lake Harbour is today. Later in the day, "one native came on board and reported that it was a topsail schooner ... and that the captain had a very long nose." These facts made it "very evident" that the marauding vessel was the *Abbie Bradford*.

In fact, Captain Spicer had more discoveries in store. The stories the Inuit told caused Spicer to contend that they had been told he was dead. And besides the *Abbie Bradford*, a second vessel, the *George and Mary*, had participated in the theft of Spicer's whales. Both were from Spicer's rival port, New Bedford.

It seems likely that the captain considered the possibility of legal redress on the spot – he gathered available evidence. Soon it turned out that luck was on Spicer's side: the *George and Mary* had carried impressive witnesses.

By chance aboard the *George and Mary* were Lieutenant Frederick Schwatka, leader of the Schwatka expedition in search of the lost Franklin expedition, and the journalist Colonel William Gilder, author of *Schwatka's Search*, one of journalism's great epics. After an overland journey of almost four thousand miles they were taking passage home from Hudson Bay.

Inuit recall that a wrecked vessel, "a schooner with two masts," figured in the thefts. This was the *Abbie Bradford*, jammed in the ice off Big Island, a customary Hudson Strait rendezvous for Inuit and whalers a few days' sail into the strait. *Schwatka's Search* supplies some details: "The *Abby Bradford* suffered such a severe strain that her timbers creaked and groaned terribly, and her deck planks were bowed up. So imminent did their peril appear that the boats and provisions were got out upon the ice preparatory to abandoning the vessel, when just as it seemed as if she must succumb, the pressure was relaxed and the crew returned to the ship."[9] But whalers and Inuit had been in close contact while awaiting disaster. "They got the whales because the schooner was old and coming apart," Pauta Saila reported. "The wood was cracking up on the bottom and it started filling with water, so it had to be abandoned. ... " It was when the crew returned to the vessel that two whales' worth of oil and baleen was taken. Witnesses heard Johnnibo say that the whales were taken against his "personal wishes" and without "permission."[10]

But it is the theft of the third whale that rivets attention. Schwatka and Gilder were eyewitnesses (Gilder was called upon to interpret), and the sworn testimony that both gave the court is oral history at its most reliable.

As the *George and Mary* sailed out of Hudson Bay, her master, Captain Michael Baker, learned from incoming vessels of the two whales taken by the *Abbie Bradford* from the Inuit at Big Island. He learned, too, that one whale still remained. The captain pushed his vessel with all possible speed. "We spoke several times of the necessity of hurrying ... in order to get ahead of any other vessel that might come in and trade with the natives ... " Gilder recalled. "It was considered improbable that Captain Spicer would come in. We thought he had returned to the United States."

Once in the vicinity of Big Island, the captain sent out a crew in a whaleboat to look for Johnnibo. "They met John Bull and some other natives coming towards the ship and returned with him," Gilder recounted. "When they came on board the vessel, we asked John Bull where that bone was. He said it was at Arkolear [Akuliak], a place further up the Bay. We asked him to trade the bone with us. He said that he could not because it belonged to Captain Spicer. We then tried to induce him to part with it by offering him articles of trade, but he still refused upon the same ground and went away on shore after nightfall, knowing that our vessel would stay there overnight. We told him that Capt. Spicer would not come back there that

The George and Mary.
The Whaling Museum, New Bedford, Mass.

When the Whalers Were Up North

Bark George and Mary.

season – that we heard from the other white men that Capt. Spicer had killed plenty of whales in the fall and had gone home. In the morning he returned with the other natives and we again spoke of the subject of trading for the bone. We told him there was no use of his keeping it for Capt. Spicer, that he would not be back. John Bull and his wife both insisted that Capt. Spicer would come back, that he had told them he would. After further conversation of the same nature, John Bull held a conversation with his friends among the natives and asked me if Captain Baker would give a boat. Captain Baker said he would ..."[11]

Lieutenant Schwatka was also present. "The argument was used with John Bull that Capt. Spicer would not be able to return ... The last conversation which was going on when I entered the cabin consisted of John Bull turning to the extreme man on his left – there were about six – and asking him as near as I could interpret, what he thought of the situation and the proposal. This native who I believe was a boat steerer in one of John Bull's boats nodded an assent. This continued down the line to John Bull himself, all of them nodding assent to John Bull's proposal or question whatever it was. John Bull then arose, saying as near as I can remember – 'all right.' "

The witnesses remembered that Johnnibo reasoned that if Spicer had come, the bone would have been his; since he hadn't arrived, probably the bone was still his own. And Annie Kimilu, Johnnibo's wife, was on board and interjected a comment. "If Capt. Spicer comes in he will cut our throats."

The *Era* log details Spicer's actions when, within days of the departure of the *George and Mary*, he arrived "to find nothing," as Inuit reported. On 27 August a scribe writes, "The Captain brought 3 natives on board to take down their statements in writing concerning the taking away of the Whale Bone & Blubber from the Station the Captain not being able to find John Bull as he was a long ways inland Deer Hunting thinking from the reports of the other vessels that the Era would not come here this year." A few days later Spicer went to the place where the whales were cut in and measured their jaw-bones. He put a whaleboat to be paid to Johnnibo for services rendered in care of another Inuk (although just in time, says an old newspaper report, Johnnibo returned from hunting) and sailed out of the strait to attend to his east-coast stations. From there he sailed home and, on 24 November 1880, at 8 p.m., saw Montauk light. "At 11.45 p.m.," reads the log, "come to anchor at New London. So ends this voyage."

Spicer wasted no time. Both the offending vessels were out of New Bedford and owned by a consortium headed by Jonathan Bourne of that port. As soon as possible after reaching New London, Spicer travelled to New Bedford and sought out the master of the

Photograph of Captain John Orrin Spicer of Groton, Conn., in a frame surmounted by walrus tusks.
Mystic Seaport Museum, Mystic, Conn.

When the Whalers Were Up North

When the Whalers Were Up North

George and Mary. "How was I to know it was your bone?" asked Captain Michael Baker. Because the Inuit had told him so, Spicer replied. "I think my word is as good as an Indian's and I say, they didn't," Baker declared. "I should take the Indian's a good deal the quickest!" countered Spicer. "Then you may as well say that I lie," said Baker. "I told him I did not say that he lied," Spicer recounted, "but I did not think anyone who would take another man's bone was any too good to do it."[12]

Spicer had no better luck with shipowner Jonathan Bourne when he went to see him in his offices. Bourne told him, "We don't back down here. We have a long purse. It is your privilege to go to the courts."[13]

C.A. Williams & Co. had no hesitation. They hired top legal aid to argue the merits of their case in court and, the following year, brought Johnnibo and his family to New England to help them seek redress.

II. THE COURT CASE

Johnnibo, his wife Annie Kimilu, her little daughter Kudlarjuk, vital as if they had posed for their portraits yesterday, can be seen in the famous 1881 photograph taken in a New London studio. A faded caption reads: "Witnesses with Spicer."[14]

Years after she was caught by the camera Kudlar-

Charles Francis Hall with Joe Ebierbing and Hannah Tookoolito, early Inuit visitors to New England, as illustrated in Hall's book Life with the Esquimaux.

juk told her own daughter her first impressions of New London. "There were so many lights."[15] The family stayed with Captain Spicer and his wife. Their appearance in their skin clothing on the streets of New London and nearby Groton where the captain lived caused comment in the press: Kudlarjuk was bright, only five but looked like seven. Her American name was Buckshot, after a deceased grandfather. Johnnibo was forty-two; his Inuit name was Chimoackjo. Annie Kimilu was thirty.[16] Perhaps she was a little more; when the explorer Charles Francis Hall met her in 1860 – "an interesting girl" – he thought she was about sixteen.[17] Many have attested to her vivacity and appeal. When more than sixty she still impressed the young missionary A.L. Fleming, later bishop of the Arctic. "She must have been a very attractive young woman," he noted in his autobiography.[18] And in old age Annie Kimilu had words to say on the matter herself. "All the captains wanted me," she used to tell relatives. "I've got worse, but I really used to be the best-looking woman."[19]

Annie Kimilu's descendants say that Kudlarjuk was her daughter by a Captain Walker, probably a Scot, since Scottish records mention a number of captains by that name. Throughout the whaling years children were born from such liaisons, but, a woman explained, "I don't think anybody minded in those days. And when the child grew up he would become a helper."[20]

The case opened 6 June 1882 in the United States

Circuit Court of Boston. The plaintiffs brought two suits of tort, one for damages of $10,000 for theft by the *George and Mary* of three hundred barrels of blubber and twenty-two hundred pounds of whalebone, and one for $15,000 for theft of three hundred barrels of blubber and thirty-two hundred pounds of whalebone by the *Abbie Bradford*, in all a total of $25,000.[21]

Both sides had retained imposing counsel. Acting for the plaintiffs, C.A. Williams and Co., were Crapo, Clifford and Clifford. The attorney William Wallace Crapo was a state senator, and Charles W. and Walter Clifford were sons of a former governor of Massachusetts. A well-known advocate, Thomas M. Stetson, was retained by the defence.

Johnnibo was ill during part of his stay in Groton, and there were fears that he could not testify; his deposition was taken, but when the trial opened, Johnnibo was in the witness box. A Captain Keeney of New London acted as interpreter. The legal proceedings were well documented, particularly in the *Whalemen's Shipping List and Merchants' Transcript*, the New Bedford whalers' trade paper, published every Tuesday morning. A wide audience followed the case with its cast from the New England whaling and legal aristocracies and its star witness – Johnnibo.

Journalist R.B. Wall, who wrote a series of articles on Spicer's life, reported that Johnnibo, through an interpreter, gave the most damaging evidence. "He acted like a trained witness, being cool and collected, brief and judicious. No witness made a better impression on the court."[22]

The crux of the conflict lay in whether or not a contract had been in force between Spicer and Johnnibo. Phillip F. Purrington, an authority on the whaling days with the Old Dartmouth Historical Society Whaling Museum in New Bedford, made a study of the case and wrote it up in 1959 as an issue of the museum's *Bulletin* expressly for members of the bar: "Counsellor Stetson for the defense denied that there had been a contract between Captain Spicer and John Bull or that the latter had been hired, alleging that it was *lex loci* for whaling masters to furnish natives with the means to catch whales, expecting only that preference be shown the outfitter in sale of the product. The fruits of the whalehunt would then belong to John Bull to sell where he wished … " The defence then suggested that Eskimos were not of sufficient intelligence to make contracts, that Johnnibo did not have the legal capacity to enter into a contract. (Normally children and "the feeble of spirit" – the insane, the retarded – do not have such capacity.)

On Baffin Island the insult is still resented and Spicer's reply remembered. Former special constable James Akavak says, "The people who owned the oil came back to find nothing, so they took Johnnibo and Annie Kimilu down south to a court hearing to negotiate. At that meeting they discussed the stolen oil, trying to understand how the whole business started.

"Whalers from a different port had come by and said, 'They – the true owners – are not going to re-

A Whaler's Dream. Pudlo Pudlat, Cape Dorset. Stonecut, 1987.

When the Whalers Were Up North

A Whaler's Dream   *Stonecut & stencil*   *Pudlo*   *1987*   *Pudlo*

turn!' So, therefore, they would take the oil. The people of the camp did not want to give up the oil, but they had no way to defend the oil from those who wanted it.

"Johnnibo was on the side of those whose oil had been stolen and opposed to the other party. The thieves told the hearing, 'The Inuit are children – just like puppies.' But the whaler whose whales were stolen said, 'No, they are not like dogs. Dogs are always stealing what they are not supposed to take. You – you are worse than dogs.' Those who were robbed liked the Inuit; the thieves didn't care much for the Inuit."

In charging the jury, the judge stressed that the essential question was what contract if any was made with Johnnibo and whether both he and Spicer understood its terms. The jury is master of the facts; the judge is master of the law, and in preparing his charge it is likely that the judge bore in mind recent decisions. In the later nineteenth century the question of native property rights was still a lively issue in the United States. A major series of decisions from the United States Supreme Court – the so-called Marshall Court decisions – had been handed down between 1810 and 1835 and constituted precedent in many cases concerned with native rights. The court had ruled that Indians had property rights and a limited degree of sovereignty.[23]

The jury in Boston found for Spicer and the consortium of C.A. Williams. It decided that a contract had existed between Captain Spicer and Johnnibo and

awarded damages in the two suits of $6,712.45.

As soon as possible, Johnnibo and Annie Kimilu sailed for home, taking with them a three-day-old baby daughter they called American Girl.[24] Spicer had tried to interest his Inuit guests in staying in New England and had provided a house and an acre or two of garden and a boat and a dog. Mrs Spicer had taught Annie Kimilu to cook New England style. But Johnnibo and Annie Kimilu said they felt much too crowded in the Groton countryside and sailed back on the *Era*.[25]

But their story is not ended. Spicer had won his case, but Johnnibo was to lose his life before a tribunal of his own people.

"He was killed. By the Inuit. Everybody knows that," said an informant flatly.

III. JOHNNIBO'S DEATH

Several years after Johnnibo and Annie Kimilu left New London, the readers who had followed the case of *C.A. Williams* v. *Jonathan Bourne* received astonishing news. On 8 January 1889 they picked up their regular Tuesday morning copy of the *Whalemen's Shipping List and Merchants' Transcript* and read that the good witness John Bull had been murdered.

The facts were reported to the press by James Jordan, an officer on the *Era* who had just returned home from the Hudson Strait. "Two weeks before the tragedy his wife was notified according to the custom, of his coming death at the hands of his countrymen, but on pain of death could not acquaint him with the information.

"When the day arrived he was decoyed out of the village by two companions and arriving at the place designated he was prostrated on his back; then the leader delivered a short speech and called on the man selected to advance and stab him through the heart while he opened the victim's clothing and pointed out the place.

"Officer Jordan says the murderer often visited the ships at anchor and was very careful to avoid the presence of Capt. Spicer, who he knew would be sure to wreak summary vengeance on him were he to know the circumstances of his faithful friend John Bull."

When Johnnibo returned from the south, he had quickly become involved once more with the whaling operations at Spicer's Harbour. But he encountered changes. C.A. Williams and Co. now had an old vessel, the *Roswell King*, stationed there as a floating whaling station. She had a white crew aboard who, with Inuit hunters, worked up cargoes that the *Era* collected on her rounds. "The Americans had their station on shipboard," says Lake Harbour's early RCMP special constable James Akavak. "In winter their post was a stranded ship. They lived all winter on their ship, frozen in the ice. In winter that's where they were, and in springtime as soon as they got a chance to break away they'd start travelling."

The vessel's log makes many mentions of Johnnibo's name.[26] It records his visits, his departures for caribou hunting. But frequently entries mention Alainga (in the log spelled Aliner), a leader of a faction that it seems, began to oppose Johnnibo actively. "Some hunters don't have trouble getting animals and don't have trouble getting women. Alainga was one of these men," says Osuitok Ipeelee of Cape Dorset. "Alainga was a happy and famous man and his great grandchildren in Iqaluit are happy and famous, too." But Johnnibo also was a powerful man. "Johnnibo didn't have trouble getting animals, either," Osuitok says. Perhaps Spicer had foreseen difficulties when he tried to keep his friend down south. "Johnnibo was killed over a quarrel about the whalers," say relatives today.[27]

Years later, Archibald Fleming, first Anglican bishop of the Arctic, heard a version of the conflict from Annie Kimilu, who, according to custom, had been taken as a wife by Alainga. In his autobiography, *Archibald the Arctic*, he reported that the hunters had complained that the mate told the whaling captain more than he should and forced them to trade everything with one man. "The hunters were very angry and warned Matte [the mate] that unless he refrained from giving information to the American captain, they would kill him."[28]

Johnnibo's fate is still discussed and debated. Many think that jealousy played a role. "Johnnibo was taken very good care of by the Americans – they gave him everything he needed – all the gear. That's probably why he was killed."[29]

Some say the handgun he carried was an aggravation. "The qallunaat had given him a handgun so

When the Whalers Were Up North

he could protect himself in case he was attacked. He'd been pretending to scare people, though Johnnibo never would have hit another human. But because he carried a handgun, people pretended to be afraid."[30]

And they say Johnnibo was feared because people suspected he was a shaman – an evil one, who used his powers for self-enrichment.

"Why was Johnnibo killed?" asks Pauta Saila. "People long ago used to be jealous of those who were treated kindly by the qallunaat. Johnnibo was rich without seeming to work hard, and the white people treated him so kindly. They were jealous of Johnnibo and accused him of being an angakkuk. And Johnnibo said, 'I have never been a shaman, but if I were I would never use the powers for my own benefit. But if you think I'm an angakkuk, why don't you kill me?'"

Agee Temela of Lake Harbour heard the details of Johnnibo's death from her whaler relatives. "Around Iqaqtilik there is a small island which used to be a lookout for whales. It was close to the floe edge and close to other islands, a good place for a lookout. It was on this island that Johnnibo was killed. This lookout was used by the whale hunters during the spring when the ice was breaking up. It's in the month of May when there is still some ice that the whales migrate, and from the lookout the hunters would see the whales in the open water.

Perils of earlier days are evoked in this coloured-pencil drawing by Jessie Oonark, Baker Lake. Collection of Mr and Mrs Brian Kaplan.

"Johnnibo was at the lookout, just about to go home, when he was killed by the Inuit. By his own people."

Among Johnnibo's heirs was a son, Isaccie. In 1987 in Iqaluit, Isaccie's son Johnnyboo Ulluakulluq said that as a boy his father had vowed to grow up and take revenge. "But when he grew up, he never did."

After Johnnibo's death John Spicer seems never to have seen Annie Kimilu again. But he did see Kudlarjuk, who told him the story of Johnnibo's murder. "He knew too much," she told him.[31]

Reflecting on Johnnibo's murder recently, Annie Manning, an Inuit schoolteacher and the first woman appointed justice of the peace on Baffin Island, came to a similar conclusion. "His death was something that could happen in those days," she said. "Inuit people were very isolated; they had many fears. It was hard for them to understand Johnnibo because he was so different. He was well off and he had been to the south. In the old days it was hard to accept a person like him. Johnnibo was an achiever."

Canada's Inuit today are stalwart fighters for native rights. "When I am gone, I have two sons at home who will follow me and speak out," one Inuit leader told former Prime Minister Pierre Elliott Trudeau at a constitutional conference.

No one can doubt it. More than one hundred years ago, many miles from home, Johnnibo stood up to defend his right and capacity to make a contract before a jury in a court of law.

When the Whalers Were Up North

Spicer's Harbour

For years in summer hunters have been travelling over the island of Iqaqtilik (on maps called Spicer's Island, but not to be confused with the islands up in Foxe Channel, also called after the captain) and finding big iron pots, whisky bottles, and sometimes a lonely grave, all relics of the days when the Americans were a presence there. Yet southern students of the whaling days have never known much about the Americans at Spicer's Harbour.[1]

Iqaqtilik is the largest island in Spicer's Harbour, a stretch of coastline that one entry in an *Era* log indicates was also known as Seven Island Harbour.[2] Spicer always spoke of his station as at Akuliak, at the harbour's western end, where Inuit say vessels anchored in the summertime, but it is with Iqaqtilik or Spicer's Island, an hour's sail away, that Inuit associate American whalers.

Crew with whaleboat collecting fresh water.
Dundee Museums.

It has been thought that vessels rarely stayed over on the Hudson Strait, but Inuit say otherwise. In the shelter of the island, Inuit report, whaling vessels regularly wintered. It was here, too, according to reports, that Spicer berthed the *Roswell King* as a permanent floating station – her log calls her anchorage King Harbour – when in 1881 she was left "clean" following some trouble off the North Bluff coast of Big Island.

Lake Harbour's premier historian is Kowjakuluk, born, he thinks, about 1908, who regularly visits the school to tell stories of the Inuit past. A widower, Kowjakuluk for many years suffered bad health, but antibiotics and care from the community nurse have restored him to health. He lives alone, loves to talk, and talks well. He convulsed interpreter Johnny Manning and me with stories of how, as a young boy, he was let down the cliffs of Coats Island, where his family lived at the time, to gather akpak eggs – eggs of the thick-billed murre. When his bucket was full, his father would pull him up, covered with akpak droppings. "But that bird shit was a good detergent. When they washed out the sealskin clothing, it came up good as new."

Kowjakuluk has done his own research on the whaling days and the people he calls "the first qallunaat." He has heard that Iqaqtilik usually had three or more vessels in winter and that they came from different ports and were sometimes of different nationalities. "When I was growing up I used to ask all kinds of questions about the whalers. That's how I know a bit about them now." There is much to relate, and Kowjakuluk talks at staccato pace. But Kowjakuluk is a careful historian. He says, "I try to tell what is exactly true."

In our talks Kowjakuluk passed on reports he received from those who observed first hand how the qallunaat whalers spent the year. The wintering practices Kowjakuluk described were much the same in all the winter harbours. About the end of September the captains would take their vessels to harbours, usually among islands, and late in October ice would freeze the vessels in. According to whaling authority W. Gillies Ross, "In strong autumn winds the whalers usually rode to two anchors, but as the young ice grew thicker, they took one up at the bow. When the ice was strong enough to endure the force of strong winds, a channel was cut forward from the bow, the ship was heaved forward (aligned in a north-south direction), and the remaining anchor was taken up. From then until spring the vessel was held by the ice alone." Thick banks of snow were built around the vessel for insulation, and much of the deck was housed in lumber brought up from the south for the purpose and lined with canvas. All extra provisions were stored ashore to make more room for living, and because of the danger of fire from the hot stoves burning aboard, a fire hole was kept open near the vessel.[3]

Provisioning the vessel with water was always a major consideration. In our interview Kowjakuluk described how the whalers at Spicer's Harbour collected their water before the onset of winter. Later, another informant described laying in a supply of ice blocks in winter. "They had long sleds and they used these to fetch ice in winter. The qallunaat – the white men – would put on a harness and drag the sled with the ice on top near to the ship, and then they would start unloading the sled and pile the ice pieces together near the ship. On top of the deck there would usually be a little building or booth with a stove inside, and in this little shack they would melt the ice. They'd use a big bucket or pot for melting.

"They used to leave the ice blocks on the shore or right on the ice, and people would carry them up to the ship. They didn't need ladders. Once there was a lot of snow around the ship, the weight of the ship would cause it to break the surface where it was, and

Whaleboat Hunters. Keeleemeeoomee Samualie, Cape Dorset. Stonecut, 1974. This artist's grandfather was an Inuit whaler known as King Atchiak; her father, Kingwatsiak, went to Scotland on a whaler.

Spicer's Harbour

the level of the ice and the ship would be the same. So it was quite easy to get the ice blocks up on the ship."[4]

When spring arrived, the whalers would hasten their vessels' release from the ice with large saws, or with ashes, salt, or sometimes gunpowder. "I have heard that Iqaqtilik used to have lots and lots of whaling ships because way way back there used to be many whales," Kowjakuluk related. The ships used to be anchored all winter long. The reason why they were in the land of the Inuit was just for the whales. So they could go whaling in the springtime. The ships didn't have motors then, and that's why they wintered over; they couldn't go as fast as those with motors. They used to whale through the whole summer and stay too late to get back home.

"Definitely the Americans were first up here with the Inuit, and the British whalers came because they heard this was a very good whaling area. I've heard that at Iqaqtilik there used to be whalers from many ports. They'd anchor their ships just before freeze-up. The qallunaat had heard of the point of Akuliak – and they'd make for it. It used to be a popular place with the Inuit for fall camping – it means the middle, 'bridge of the nose.' From Akuliak you can see Iqaqtilik, and just before winter came the qallunaat would anchor their sailing ships at Iqaqtilik.

Affectionate Mother. Pitaloosie Saila, Cape Dorset. Stonecut, 1985.

When the Whalers Were Up North

"The Americans had no real community there, but it seems they had something that looked like a dry dock where the ships were pulled up on the land for maintenance. This dry dock was in a cove where it was difficult for big boats to come in and out. In the cove there was very little room because of all the pressure ice.

"I don't know how much the Americans were using Inuit helpers, but there was an Inuit camp at Iqaqtilik – the word has a meaning connected with drying meat. They used to make a year's supply of dried meat at that place. At Christmas time the Inuit used to go on the ship to feast and also probably to play games. They used to play games both indoors and out, but I don't know what kind of games they played.

"Just before freeze-up the ships lined up one in front of the other in the direction in which they would start to move when it was time to go out in the spring. They weren't side by side.

"The white men got their water supply from an island that wasn't that far from Iqaqtilik. The island was rocky and steep, and the ships used to line up near that island to get their water – near Kanisilik, where the pack-ice doesn't build up. They'd have to go over the ice and get their water from the pools on the hangio – the ice on the land before you get to the

Man Harnessing Dog. Pitaloosie Saila, Cape Dorset. Stonecut, 1985.

pack-ice. Some of the qallunaat were quite slow. A lot of them would line up to haul water to their ship. They'd move it to the ship by human power, and some of the crew would get quite slow, like dogs. Humans can be as slow as dogs. When one ship had its water supply, the next day they'd start on another.

"The whalers used to use their sails to cover the top of the ship, their cabins, and the entrances, to keep the cold out and the snow from going in. That's what those who lived before me said: that the whalers used their sails as shelter. Nothing else – there were no actual tents. They used strictly sailcloth. I've heard that there'd be two to four masts, and although they wondered, nobody ever knew whether there were whaling ships with six masts. Some were short and some were tall.

"There was once a ship at Iqaqtilik that suffered a very bad fire, and because of the damage it sank. The people before I was born, probably my great-grand-parents, knew of the fire. The crew tried to put out the fire with water, but there were no pails. And I have heard there were watches on duty even when the fire was so bad it was out of control. The ship-wrecked crew was lucky because there was another ship not too far off and the people moved to the other ship. Those who wanted went out with the Inuit hunters. We were nomadic in those days, so they probably spread around the different camps. They ate raw meat and lived the Inuit way of life. A lot of them got used to hunting and eating country food and just

when they really got to like it, they had to leave. They left when the ships were coming up in the spring.

"The ships would go out in the spring when the mouth of the cove would be open and relatively free from ice. This would be a little bit earlier than inside the cove because of the strong currents. These helped to keep the entrance free of ice, and the whalers would go in and out as they pleased in the springtime. First they would use massive saws to free the ship from ice. The anchor would be on the ship and not under the water, and therefore the ship would be free from the ice because the ice would be removed by using the saw all around the ship. The ship was not an-chored, but there were ropes going to the land to keep it from drifting with the current, to keep it in one place. It would seem to the people who had not been to the ship that the ship was frozen right into the ice, but apparently it was floating in the water, according to the people who actually went to see.

"In summertime there would be whalers busy whaling in the area. Around Iqaqtilik and in the vi-cinity they caught so many whales they could not take all of what they caught back to their homeland. The ship stayed around Iqaqtilik and the whaleboats would stay near the land and scout around towards Itiniq because the waters there are very deep and good whaling waters. There is an island which they called a lookout point, and the whalers would go to this island and observe from the top with a telescope. And also I happen to know that there was a place round here called "arvinavik" – whaling station. Here they would cut up what they caught, especially the blubber and baleen from the whale and the blubber also from the walrus and the beluga.

"They would do this year after year.

"That's the information according to the people who were there to witness the happenings. I'm sure there would have been more from people older than I am, but now none are living to tell."

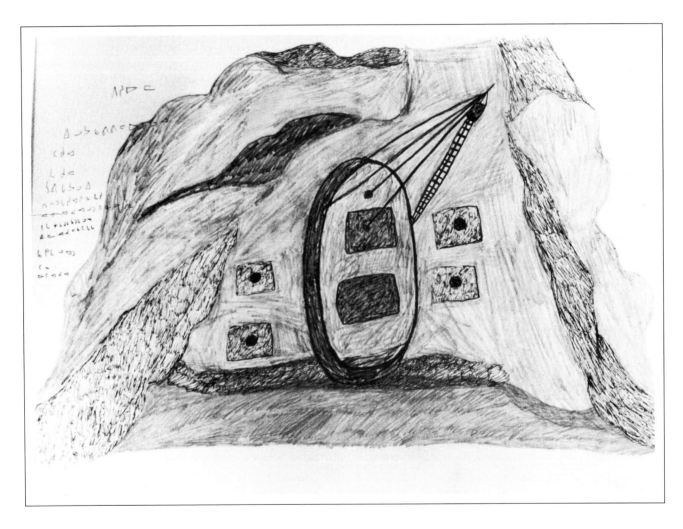

When the Whalers Were Up North

The Siikatsi and the Wreck of the Polar Star

"The ships I remember hearing of were the *Active*, the *Polar Star*, and the ship of Mitsiga[1] – Mr Grant," recalled Pitseolak Ashoona of Cape Dorset. "People used to mention these ships because they travelled around this area. My mother was always talking about those ships. I think they were whaling ships; the Inuit used to be whalers."

This colour-pencil and felt-pen drawing shows the wreck of the Dundee steam whaler Polar Star. *The vessel was wrecked on the coast of South Baffin Island in 1898, some ten years before the artist Pitseolak Ashoona's birth. Beside the wreck she shows the large metal vats the whalers brought out stuffed with provisions and took home filled with oil. Pitseolak made this drawing – one of three depicting the* Polar Star *– in 1981, two years before her death. In her syllabic note her fatigue is evident: "The steel bases used to be used as a foundation for houses. I did not sketch very well for I have been quite ill. The ship should have been more in at an angle but at times I can't sketch properly. The ship is docked too far on the beach."*

Pitseolak, who died in 1983, was a noted artist justly famous as an interpreter and illustrator of the old Inuit ways and customs. Her pictures are filled with motion and action. "Walking, jumping, running – I've even done skipping," she told me when we did our last interview together not long before her death.

The ships Pitseolak spoke of were the vessels of the Siikatsi – the Scots – new people who appeared in the strait and Hudson Bay in the last years of the nineteenth century.[2] Lake Harbour's historian Kowjakuluk says that after Johnnibo's death the American whalers returned to their Hudson Strait harbour only one more time. Activity at Spicer's Harbour subsided. He sometimes wonders, as perhaps others did at the time, if Johnnibo's murder was the cause. "They came back after Johnnibo was killed and missed him. He had been looking after their oil. After that it seems that they didn't come back again. I myself have thought that perhaps the reason the whalers didn't return the next springtime was because Johnnibo was murdered. I never asked anyone about this – it was just my feeling."

But in fact by the late 1880s eastern Arctic whaling in the strictest sense was in decline. Fewer whales were being caught and fewer whalers were pursuing, or, in the parlance of the time, "persecuting" them. "The whaling is done; we are sinking money every year," pronounced Captain E.B. Fisher, a veteran of the Bay, in 1897. "The whales are easily disturbed and they leave the grounds; they are also being killed faster than they increase. We used to get some very large whales in the Welcome, but not now; this was thirty years ago."[3] And Captain Thomas McKenzie declared, "Whales are getting scarcer every year."[4]

The vessels that came through the strait in the late 1890s were hunter-traders as much as whalers, after furs as much as oil or baleen. Captain Spicer sold up operations of the Williams firm at Blacklead Island probably in 1894[5] and seems to have discontinued operations at the Hudson Strait station in 1889. The captain retired and spent most of the rest of his life on his farm at Groton with the jaw-bone of one of the largest whales ever caught as an archway at his garden gate. (A great nephew in Groton, where the Spicer family still lives, remembers in his childhood the captain helping him to build a whaling ship from blocks.)[6] But the *Era* still sailed, under a new owner, Thomas Luce & Co., and with a new master, George Comer, now the sometimes solitary but always dominant American presence in Hudson Bay. He took the whales if he could get them, but he also wanted muskoxen furs, bear, and fox.

Comer had made his first voyage under Captain Spicer and was both skilled and resourceful; in depleted waters he still made whaling pay. But Comer had broader interests and over his Arctic career made important contributions to cartography, ornithology, and ethnography. He worked closely with Franz Boas, the "father" of anthropology, and helped to build up Inuit collections for both European and American museums. And, using cameras consistently over a period of years, he compiled a remarkable photographic record of whaling and Inuit life.

In the late 1890s Comer found he was sharing the whaling waters with the Scots. "We used to call them the people with the loose tongues – because of the brrr in the voice," says Pauta Saila.

As the Americans had done, the Siikatsi established close ties with the Inuit whalers around Big Island. Late in the 1890s the Tay Whale Fishing Company of Dundee began to send vessels to the Bay, stopping by Big Island to pick up Inuit crews and employing as officers the Murray brothers, Alexander and John, whose father Alexander Murray Sr, also a whaling master, had participated in the Franklin searches and received a commemorative medal.[7]

The Polar Star. *Colour-pencil and felt-pen drawing by Pitseolak Ashoona, 1981.*

When the Whalers Were Up North

The Siikatsi and the Polar Star   71

The Murrays were popular men, known, Inuit say, "all over the North." Both first went to sea with their father, wintering at youthful ages in Cumberland Sound. In a career filled with incident, John Murray, born in 1863, voyaged in the Arctic on many vessels until 1932. He lost a leg on a 1919 voyage, and captained the Hudson's Bay Company supply ship the *Nascopie* from 1928 to 1930. Alexander Murray's career was shorter and, as Inuit will relate, its last days were clouded by tragedy. The Murrays were familiar with the Bay since Alexander for two voyages and John for one had captained the *Perseverance* when the HBC, earlier in the decade, made a late, brief foray into whaling.

The Tay Whale Fishing Company, backed by wealthy jute manufacturers who needed whale-oil for manufacturing purposes, was under the direction of the particularly energetic Robert Kinnes, who also set up two partnerships, the Cumberland Gulf Trading Co., active in the Cumberland Sound in the years after 1905, and the Robert Kinnes Company, which as Kinnes Shipping Ltd., established originally it is claimed in 1883, is still active today as shipbrokers, shipping and forwarding agents serving North Sea oil operations. Kinnes and his companies are remembered as perhaps the most business-like of the so-called Scotch free traders (essentially recycled whalers with an interest in furs as well as whales) who, as more and more whalers abandoned the chase, began in the early twentieth century to operate in the eastern Arctic.

They used chiefly old vessels – but steam-whalers – from the once-powerful Scottish whaling fleet, formerly most active in the Greenland Sea and Davis Strait and Lancaster Sound. Their principals bought up the old whaling shore stations and also established new ones. Cape Haven at the ancient camp site of Singaijaq, certainly one of the oldest rendezvous points of whalers and Inuit, fell into the hands of "Mitsiga" – Osbert Clare Forsyth-Grant, one of the more flamboyant free traders, whose exploits engendered stories still circulating among South Baffin Inuit.

For Inuit in the strait, the Siikatsi era opened in 1899 with a bonanza – the wreck of the Dundee vessel *Polar Star*. Her captain abandoned ship on the coast of south Baffin Island, running her ashore at a point called Itilliajjuk, where, it is remembered, she was "left for the Inuit." Her crew was taken off by the *Active*.

In our talk Pitseolak, who was born in 1908, made special mention of the joyful salvage operations that went on for years and drew pictures to illustrate activities she remembered vividly. "It seems there was no wood around this area until the *Polar Star* was

Pitseolak Ashoona shows how each summer when she was a child her family and other families would camp near the wreck of the Polar Star *and "take apart the ship." She explains, "We had no wood up here until the* Polar Star *was wrecked." In her syllabic note she writes, "During the early fall the tide washes kelp and other things up on the shore. But I am tired again so I finished it." Coloured-pencil and felt-pen drawing, 1981.*

The Siikatsi and the Polar Star

wrecked at Itilliajjuk. I think the reason Mitsiga and *Polar Star* got wrecked was because they were always driving in very close to the beaches.

"I'm the only one left now to know about the *Polar Star*. We used to play on the decks of the wrecked ship when it was still whole. Yesterday I took what I had drawn to the Co-op. The picture shows lots of tents because that's how it used to be. A lot of families used to gather in Itilliajjuk to take apart the ship and get some wood during the summertime. That's when we started using wood for sleds, kayaks, and tent poles. I was on the ship when I was a little girl with my father and his brothers. We children used to play with the glass that was used for the windows on the ship. We used to have the portholes as our toys, and we liked them very much. Because it was a wooden ship, it was useful for so many things. The Inuit used up all the wood. They used up a whole ship."

Summer after summer Inuit took away wood from the wreck. Its presence changed their culture. Kayaks and qamutiit – sleds – from the salvaged wood were in use for decades.[8] A Lake Harbour hunter remembers he once had a kayak made from that wood that had belonged to his grandfather. "Usually they changed the skins on a kayak every two years, but on that particular kayak the skins were changed every year. The frame was so valuable because there was no wood available. They had to get it from the wrecked boat. In earlier times they used driftwood. Even if a piece of wood was very small, they could attach another piece. The most difficult part was to find a piece that was naturally curved for the curve in front of the seat. That was difficult to make; they just had to search until they found a piece that was suitable, that had the right curve."[9]

After the wood of the *Polar Star* was all used up, the iron barrels in which the crew stored the blubber still remained. Lake Harbour's historian Kowjakuluk says, "Some of those big square barrels are still out on the land near where the wreck used to be. I've heard some Inuit used to make qamutik runners out of the metal.

"The whalers had used those barrels to store their oil. But before they put in the blubber, they had to get rid of their biscuits. The first qallunaat didn't have flour – they had biscuits. The big square barrels with the small openings were first full of biscuits, and then, when they ran out of biscuits, they filled them up with oil. Most of the barrels on the *Polar Star* were filled with oil, but some still had their biscuits."

Eventually some of the barrels were transported to Cape Dorset, where people still point them out to you today. The rest remained at Itilliajjuk. One harsh winter, after the Inuit had become Christians, a man in a family camping nearby died. It was too difficult to gather the stones to cover the body the way the missionaries had taught, so his relatives placed him inside one of the big iron barrels. People felt that this was not a comfortable resting place, and recently, after many decades, they took him out and gave him another burial.

The news was on the CBC.

*"In the old days, to be a good hunter a man had to be brave,"
Pitseolak Ashoona of Cape Dorset declared. In this felt-pen
drawing, this kayaker in rough seas wears an akulisaq – a
waterproof jacket made of gut of the bearded seal and attached to
the opening of the kayak. "I knew a man who didn't put on his
akulisaq and he died," Pitseolak recalled. In her syllabic note
Pitseolak explains that if a hunter wears his akulisaq when he is
seated in the kayak no water can enter. The akulisaq is laced
tightly to the kayak aperture and the hood and sleeves are tied
tightly. About 1973.*

The Siikatsi and the Polar Star

CHAPTER 7

Death of the Last Tuniit

When Inuit looked through my pack of pictures, they often stopped and spent a long time examining a fuzzy, indistinct image showing four men in polar-bear pants, perhaps with their hair done in topknots. This was a photograph by Captain Comer taken of the isolated people of Southampton Island – the original Sadlimiut – who still used stone tools and the most primitive of weapons when they were met by the whalers in the late 1890s. They died out – only one woman and four children survived – from disease introduced by the whalers in 1902.[1]

The story of the Sadlimiut touches on the saddest aspect of the history of white contact in the North. No Arctic horror stories can match the tales of families destroyed by TB, by measles, and by other diseases against which the victims had no immunity. An el-

Two Sadlimiut children who were adopted by Aivilik Inuit and survived the epidemic that caused the death of the last Tuniit. Geraldine Moodie, 1904–05.

Museum of Mankind, British Museum, London.

derly Cape Dorset woman who remembered the construction of the Hudson's Bay Company posts at Cape Dorset and Lake Harbour said, "Every time a ship came in, the Inuit used to get colds."[2] Colds were the least of it. Pitaloosie Saila of Cape Dorset recalled that there were about one hundred Inuit suffering from tuberculosis at a Hamilton, Ontario, sanatorium when she was there as a child in the 1950s. "It was difficult for the people who died there. It used to be sad when you saw people dying away from their families." The social diseases the whalers brought also affected the birth rate. "We say today that families have too, too many children. But in those days some of the women were sterile and not able to bear children," one woman in the Keewatin told me. "The complications following [the introduction of venereal disease] were very widespread and will have a marked influence on the vital statistics for years to come," wrote Dr L.D. Livingstone in a 1925 government report.[3] Fortunately, with the advent of antibiotics, disease in the North has come under control.

When the Whalers Were Up North

Scholars still debate the identity of the Sadlimiut, the original people of Southampton Island. Some have thought they were the last survivors of the people the Inuit call Tuniit, the early people of the Dorset culture who lived on Baffin Island until about A D 1300, when they were overtaken by waves of new people, the Thule culture Eskimo arriving from the west. The Inuit whalers who met these unusual people had their own special name for them; they called them the Pujait – a word meaning "dried-up oil" – because of their extreme messiness. Also the Pujait had another odd characteristic – they spoke in "baby" voices. The Inuit whalers themselves believe the Pujait were Tuniit, stone-age people untouched by the white man's world.[4]

The isolation of the Pujait was disrupted first by Captain Comer, who in 1896 began sending whaleboats on springtime cruises to Southampton, and then in 1899, when John Murray set up and managed a shore station there for the Scottish interests.[5] Previously the Pujait had very rarely had outside contact. Crews from the *Abbie Bradford* visited them briefly in 1878 and 1879 and left a few of their fascinating trade items – beads, needles, metal knives, mirrors, and thimbles – but essentially the Pujait or last Tuniit had lived undisturbed.[6] In fact, even other Inuit groups knew very little about them. Years before, it is said,

In this coloured-pencil drawing by Jessie Oonark of Baker Lake, powerful spirits appear to direct the sea mammals.
Collection of Mr and Mrs Brian Kaplan.

South Baffin people used to go over in their skin boats to visit these people who talked in such a strange manner. South Baffin islanders called them then "Takoogatarak," meaning "we are shy with them," because before the skin boats got to shore, the men there would try to trade wives with them and say in their baby voices, "They don't want it; they don't want it." But before the whalers arrived, no one had been to them for years.[7]

Comer has described his meeting with these people: "When in 1896, I first met the natives of the island, they numbered about seventy, and as our boat approached the island near Manico Point, the men and children followed along the shore until we found a landing place. They made short, high jumps and called out in imitation of the great loon, 'Whar whee! Whar whee!' an expression which they always used to denote appreciation and pleasure. The Eskimos from the mainland on board my vessel assured me the presence of the children was an indication of good will, and as we were particularly anxious to make certain inquiries regarding the whaling prospects, we decided to go ashore. The island Eskimo … led us to their houses – seven in number and located near the coast – the first of the kind that I had ever seen, since the Eskimos of the mainland live in snow houses and skin tents …

"The island huts we found to be circular in shape, skillfully constructed of limestone and built partly underground. The roofs consisted of a framework made of the long jaw-bones of the whale, the inner ends resting on a king post; upon these bones were flat limestone blocks, and over these a layer of sod. Light was let in through an opening above the entrance, over which was drawn a piece of translucent parchment from the intestine of the seal …

"The Eskimos of my party … found it difficult to make themselves understood, for while the dialect is similar, the intonation is quite different … "

The Sadlimiut were great whalers, and Comer admired their daring. "That they were a fearless people is evidenced by the numerous head-bones of the whale which are to be found in the construction of their houses," he said. "For an Eskimo in his frail kayak to attempt to capture a whale with the primitive implements which they manufactured meant great courage, although it is probable that, in general, only small whales were taken. In the summer the natives ran out on the ice and harpooned the whales without using a boat, which, of course, involved but little danger."

Comer got to know some of the natives quite well. One was called Cumercowyer, which meant he "could see the whale under the water." Before Cumercowyer died he requested "that his body should be placed on the ice so that later it would drop into the sea. At the same time he charged his people that when they went off on the ice or in their kayaks for whales, they must throw a piece of meat into the water and call on his spirit to aid them. He promised that he would hear their call and come to their assistance. Being a friend of Cumercowyer I also was supposed to throw over a piece of meat and invoke his spirit to help us in catching our whales."[8]

In 1899 the remote and isolated Pujait suddenly had Inuit from both Hudson Strait and Hudson Bay living on their land. John Murray had with him Inuit from both areas at his new Southampton Island station. They were brought by the *Active* with his brother Alexander as captain.

Oola Kiponik of Lake Harbour remembers the stories told by the Inuit whalers who spent a whole year there hunting whales. "I think the whaling ship left people behind there for the winter and then went back for them later. My parents used to talk about the Pujait who had never seen white people before that time. They were very sloppy people, sloppy with their blubber. Because they had never seen or been influenced by the qallunaaq, they were still very messy.

"They had the same tools the Inuit had, but some of their tools were made of stones. One Inuit on the ship asked a Puja how he would aim a rifle, and this Puja pointed the rifle the wrong way round. He could have shot himself!

"They could talk to other Inuit, but they had a different dialect.

"These Pujait were found when the Inuit from South Baffin Island were on the ship *Active*. They had never met white people before, and whenever there was any commotion or activity on the whalers' ship, the

Pujait would get excited and clump together in a very close group. Everything the white man did was very new to them. They were scared.

"Because they were caked with oil and filthy they were always called Pujait – "dried-up oil". Even other Inuit called them Pujait. Even my mother. My mother herself was no longer Pujait because, very often, more frequently than those people, she had new clothing of polar bear and other skins. In the winter season my parents would have polar-bear clothing because it was thick and in spring they would try to have light clothing – sealskin or caribou, made from the skins of baby caribou, because it was thin.

"My mother used to talk to me about the Pujait women. They had their hair braided in a topknot, and because it was so puja it was so stiff that it seemed glued together. They were very careless women. And my mother talked a great deal about the men and how badly pujait they were. The men even had their hair in topknots like the women.

"Even though my people were Inuit they used to talk about 'those poor Inuit' – as if they themselves had never been like that."

Captain John Murray also left an account of his first meeting with the Pujait. It is an oral account, passed down in his family, and his son Austin Murray recounted it to me when I visited him in Wormit, Scotland, in his house overlooking the Tay, the great artery on which the whalers sailed out from Dundee. Murray lived in close proximity to the last Sadlimiut

for many months, and Peter Freuchen of the Fifth Thule Expedition, asking to borrow Murray's photographs for publication in the expedition's report, wrote that he had heard that Murray was "the expert."

Remembering his father's story of the first time he met the Pujait, Austin Murray related, "They were on a hill and kept going further back all the time. He sent up one of the Eskimos and eventually they were persuaded to come down. One touched my father's face – to see if the white would come off. Eventually he got them to come on to the ship, and he put on a Harry Lauder record. They were very subdued at first, but this was a laughing song and full of ha ha ha and eventually they took the giggles, as it were, and were laughing their heads off. When he left them they looked down the gramophone to see who was speaking. And took a bite out of the record.

"They had only bows and arrows with flint heads on the arrows.

"They got the influenza and had no resistance."

It may not be a coincidence that images of Inuit enjoying a Harry Lauder record and inspecting a gramophone speaker appear in the 1922 documentary *Nanook of the North*. Some years after his sojourn on Southampton Island, Murray met filmmaker Robert Flaherty and was well acquainted with him by the time Flaherty made his pioneering classic. In fact, for a later movie, *Man of Aran*, Murray was hired to handle the special effects.

The poor Pujait died out suddenly in 1902, and for

many years there was speculation why. Captain Comer thought that perhaps the Pujait starved because of overhunting by the Inuit and crews working for his rivals, the Scottish whaling station.[9] South Baffin people believed that the Pujait died out because of enmity between the shamans. The Pujait had two chiefs, the first chief Kamakowjuk (perhaps Comer's Cumercowyer, who saw whales under the water) and his brother, the second chief, Avalak, who astonished the ordinary Inuit with his great strength. Avalak and Pitseolak Oojuseelook, who was a powerful shaman from South Baffin Island, had a test of strength, and "Pitseolak just went down immediately and flipped over a couple of times." People thought that Pitseolak the shaman killed all the Pujait "because he lost in the strength game."[10]

But this may not have been the opinion of Aivilik hunters from Roes Welcome Sound. John Murray's Inuit boss over many years was Angutimmarik, called by the whalers Scotch Tom. He lived through the 1940s and is remembered as a great shaman with a sophisticated understanding of illness. An Inuit woman who camped with him in later years remarked, "Angutimmarik! I think everyone remembers that man. Nobody was ever to use his things. Even his cup was never to be used, but I didn't know this and I took the cup and drank from it. In our camp a couple of people were really sick, and Angutimmarik began doing his magical things to discover who was causing the illness. After using his magic, he knew I had used his cup, and he told me I was not to do this because in

Angutimmarik (left) with two Sadlimiut men. John Murray, 1899–1902. Fifth Thule Report.

this way he would pass all his germs to me."[11]

Angutimmark had his own tragedy that summer of 1902: along with the Pujait, one of his own children died. About twenty years after the death of the Sadlimiut, he met Therkel Mathiassen of Knud Rasmussen's Fifth Thule Expedition and told him that a ship had called at the Scottish station that fateful summer and brought with it disease. In a 1975 publication, *Whaling and Eskimos: Hudson Bay 1860–1915*, W. Gillies Ross showed through a study of log books how epidemic was introduced in several areas that year by the *Active*, which carried a sick seaman who died on board. Many died, but only the Sadlimiut, with little resistance to disease, were virtually wiped out.

Before the tragedy, according to the report of a member of the Fifth Thule Expedition, Angutimmarik adopted one of the four Sadlimiut children who survived.[12] Joe Curley, Angutimmarik's nephew and adopted son, spoke of his step-sister and the disease that destroyed the last Tuniit: "In earlier times Southampton Island had its own people, but when sickness struck, the people living there were wiped out. At the time we wondered what was the cause of death. My father adopted one of the orphans. She was my older sister. Her name was Nanutak Akpaliapik – she was one of the early people to survive. All her people died. No, she wasn't distinctly different. Through adoption she grew up used to our pattern of life.

"My sister Nanutak ended up marrying a man called Kamartak from Cape Dorset. She had a child called Saimanaaq who bore a daughter who was adopted. Her name is Susan Siksik, and she's married to Joe Savigattak, so she's one of that family now.

"Those people were supposed to have lived in the Stone Age. When they died they were just being introduced to weapons and modern life. They had been living in stone buildings and using stone tools.

"They died just as they were getting guns."

When the Whalers Were Up North

The Active

In treasure boxes carefully stored by families in Cape Dorset and Lake Harbour are much-cherished collections of big British pennies. Often these are inheritances passed down from whaler ancestors who travelled on the *Active*. The pennies once decorated women's amautiit, along with beads and spoons. "Only rich people had these things," the Cape Dorset artist Pudlo Pudlat told me. "Most people didn't have a spoon in the house." Older Inuit love to remember the sound the pennies made when women danced in their decorated amautiit. They say the Queen Victoria pennies made more noise than pennies with the heads of other monarchs. The reason? "There was less metal in the old Queen pennies."

Regularly for many summers the *Active* passed by the camps close to Big Island and present-day Lake Harbour and picked up Inuit for the hunting season. Under the direction of energetic Robert Kinnes, the

The Active *in port.* *Dundee Museums.*

Tay Whale Fishing Company also established a mica mine close to Lake Harbour which they worked with Inuit labour. The mica was used for stove doors and automotive batteries. "My ancestors were chiefly geologists," James Akavak told me.

In 1900 Kinnes determined to visit the Bay and, stopping near Lake Harbour, inspected the mica deposits. His journal entries describe the start of mica mining – the operation was the first mine on Baffin Island – and also, perhaps, the start of the community of Lake Harbour. The presence of the depot built by the whalers later influenced the Anglican mission and the Hudson's Bay Company to build nearby.

For Saturday, 25 August 1900, Kinnes wrote, "7.30 am found all the boxes on deck and loading boats with them, sent also on shore the wood … and instructed the natives to build a house to store the mica at the head of the fjord – took the name of those who are willing to work at it and noted what return they want for their labour if they succeed in filling all the boxes with good mica."

On 28 August he wrote, "I have no doubt if the place was properly worked good deposits would be found underneath as at every point we tried we could always get mica in quantity small and indifferent pieces to start with but improving as you went in."[1]

After the return journey, when the *Active* docked at Dundee a local newspaper reported on 23 September 1900, "Mr. Kinnes has brought home one of the natives from the settlement on holiday. Schaa or "Billy" as he is styled by the crew is most intelligent and talks English fairly well."[2] A few years later Kingwatsiak also accompanied the whalers back home – for the purpose of learning more about how to cut mica.

The *Active*'s route usually took her to Repulse Bay, where, following the death of the original Sadlimiut and the abandonment of the Southampton Island station, the Dundee whalers maintained a ketch as a year-round station. With her auxiliary engine the *Active* had no need to winter over, although Lake Harbour people remember that on two occasions she did so; usually she would act as a tender and off-load goods from the ketch for transport back to Scotland, deliver provisions, and return to Lake Harbour to pick up the mica, hunting the sea mammals along the way.

The *Active*'s captain was Alexander Murray. Inuit called him "Capitane" and remember him as "a very good captain." Lake Harbour's historian Kowjakuluk reported, "He'd had an Uumanaqjuaq childhood. A lot of white people went up to Cumberland Sound with the ships, and his father was a whaler who had

had a ship. So the captain grew up at Blacklead Island and learned from his father how to be a good captain. He was a very good man to the Inuit. He grew up in an Inuit camp and he knew how we lived. He must have spoken our language well."

Kowjakuluk had heard that Alexander Murray Jr's first command was his father's old vessel the *Windward*, apparently converted to steam at the time he took over. "The son took over his father's ship. When his father had it, it didn't have engines. After he got it, it did."

When the *Active* arrived each summer, Inuit families went aboard, taking with them "the dogs, the sleds, the umiat – the sealskin boats – the kayaks, everything they had." For many years the lives of many families revolved around the *Active*.

One of the children on board was Murray's Inuit son, Ikidluak of Lake Harbour, one of the most respected of South Baffin hunters. Some of his earliest childhood memories are of the *Active*. "I was born when the ship *Active* was coming up around here. My father was the captain of the ship, Captain Murray. I have heard that was his name though I don't remember him. I used to think that my mother's husband was my real father. His whaling name was Iyola and his baptized name was Abraham. Yes, I remember going on the *Active* – just little bits. It's as if I keep falling asleep. My memory is blurred – it has been so many years.

"I remember the front of the ship had a carving. A qallunaaq with a hat. He was standing with one foot raised. On the mast there was a barrel where someone could sit, and there was usually a person there with binoculars watching for sightings. They would look not just for whales but for walrus, seals, and square flippers, too.

"I don't remember them catching whales, but I do remember when they got some walrus. I remember the women taking the blubber off the walrus. They had a long board for the women to work at and a long line of women there taking the blubber off the skins. I don't know how many, but there were lots of women scraping. Some were just standing by to help. The women were working while the men were out catching the walrus. After the blubber was off, they would salt the hides and put the blubber in the blubber barrels.

"I don't really recall eating maktaaq from the bowhead whale, but I do remember my mother and the mother of another little boy cutting up maktaaq in small pieces for the two of us to eat. We ran with them up to the front of the ship and dropped them one by one into the water. Then we went back to our mothers and got some more to play with. So I don't remember not liking it – maybe I never tried it!

"In the mornings the cook used to bring a big pot of soup up to the top of the ship and put it in front of the Inuit. They'd use their cups to dip into it. There's no doubt we were given more than just soup, but that's all I remember – just that sometimes that big pot of soup would be brought up in the mornings carried by two people. And I remember that maybe

Figurehead of the Active *(detail).*
Dundee Museums.

twice I was called through the loudspeaker by the cook to go into the kitchen and get something to eat there.

"The Inuit used to sleep in the hold – where they kept the luggage. They would crowd in and all along the wall they would have boards running around and those boards were used for beds. They would be completely full of people. The Inuit would be sleeping there and on the floor, too.

"Twice the ship wintered over. Once I've heard

Alexander Murray Sr, Peterhead master and veteran of the Franklin searches. Courtesy of Austin Murray.

Alexander Murray Jr as a child, about 1865. Courtesy of Austin Murray.

When the Whalers Were Up North

the ship stayed over in the Repulse Bay area. The people from Lake Harbour made up a team and they played games with the people from up there. There was a team from there and a team from here. The people on the ship and the people from Repulse would try to pull each other with all their strength to see who was stronger. Tug of war."

As Lake Harbour's historian, Kowjakuluk collected many impressions of life and activity aboard the *Active* from those who voyaged on her. "I was already born when the ship the Inuit called Umiarjurapik – beautiful ship – was up around here. It used to come around Lake Harbour where the small camps were in the spring. They were whaling and hunting for walrus tusks, walrus hides, walrus fat, and for belugas and big whales. They'd be after the blubber all summer. They used to gather up Inuit so they'd have helpers aboard for guidance and everything else. The Scotsmen had Inuit as harpooners and boat steerers, and the whaleboats had only Inuit people in them. Sometimes a few qallunaat. The ship had six boats – three on either side – but there didn't seem to be enough people on board for all the boats.

"At the top of the mast was a wooden barrel – shiny wood, plain-coloured like chairs – and big enough for two people to watch – two Inuit, two Scots, or an Inuk and a qallunaaq, whoever wanted to climb up and watch. The captain was usually up there, and anybody who wanted to see him, the mate or an Inuk, would have to climb up. They looked for whales through the big telescope, and when they sighted a whale they'd put out the whaleboats and the whalers to run after it. They had a pointer, a big black arrow, to give directions. The pointer was directed to the whale. They made the pointer big and painted it very black so the whalers in the far distance could still see it – so long as there was no rain or fog. The whalers would row in that direction even though they couldn't see the whale. It was a long time ago, and they had no radios then.

"When the men on Umiarjurapik went after the whale, only the captain and the cook would be left on board. They wanted to go but there was no one left to pull up the anchor, so they had to stay there with the pointer and wait for the whalers to come back. Just the captain and the cook and the wives would be left. Everyone else would be scattered."

Particularly keen memories of the *Active* belonged to Anirnik of Cape Dorset. Born aboard the *Active* in the last century, she was pregnant with her first child the year after the whaling voyages ceased. Until she was thirteen or fourteen the *Active* was her floating home. Hers was one of the great whaling families, her father one of the greatest hunters. Each year Aningmiuq, who caught three bowhead whales for the *Active*, and his brothers Agirnik, Laisa, and Toonillie with all their wives, children, and dogs would board the vessel for the season.

When I first met Anirnik, she frequently drew dramatic mask designs for Cape Dorset's print-making program. There were shamans aboard the *Active*, and one day I asked, "Are those shaman's masks?" An-

Aningmiuq, Inuit whaling mate on the Active. *Robert Flaherty, 1913–14.*

Notman Photographic Archives, McCord Museum of Canadian History, Montreal.

irnik, a good Anglican, replied, "Maybe they are … " and then changed her mind: "But I don't think of shamans when I do them."

When we had our last talks together, I was saddened to discover that Anirnik had become almost sightless. "When my eyes were good," she said, "I could have drawn pictures of the *Active* – if I had thought of it." Her word pictures offer fair substitute.

"They used to go down past Lake Harbour and wait for the ship. Even before the ship showed up they would see the smoke and everyone would start rushing, getting the kayaks ready because they were going on the ship. They would make a lot of noise getting on the kayaks. Rushing, hurrying. The ship would blow its horn and maybe the hunters would start shooting. They used to get excited when they were going on the ship.

"Today when a ship comes in, they're not so happy. People hardly notice.

"I was born on the *Active*. When I was a little girl we were on the *Active* every year and all summer because my father was hunting bowhead whales in Arvilik – the land of big whales. When we saw the smoke we started to pack our things to be ready to board the ship. We travelled on the *Active* to the place where the tide never gets low, and that's where we'd look for the whales. I have never seen whales here near Cape Dorset, only where we went with the ship. The ship's people were Siikatsi – Scots.

Mask. Anirnik, Cape Dorset. Felt-pen drawing, 1967–75.

When the Whalers Were Up North

"When we travelled through the ice my father An-ingmiuq was up on the captain's bridge. My father and Iyola, Ikidluak's father, were acting as bosses. They had their own whaleboats – they did not belong to the *Active* – in which they went after the arvik – the big whale. The *Active* stayed anchored and the whaleboats went after the whales. They did catch whales, though not in great numbers.

"There were many men helping my father and there used to be a lot of us sleeping in the hold. Those people are no longer living today. There used to be three layers of bunks all around the hold, and for shielding you from other people they used to have sealskin curtains. There were no white people in the hold – only the Inuit. The qallunaat had their living quarters in the front and the back of the ship.

"Sometimes there'd be big storms and everyone would be seasick, but I was always running around. It used to be lots of fun – there were quite a few of us, all about the same age, and we used to run around the decks. We used to play games, and the white man never scolded. When it was stormy we used to be instructed to stay inside, and whenever we went through the ice the little ones were told to stay inside.

"We little ones used to have 'pretend' fathers. It was just a game. Because our mothers still had their original husbands. I think the crews pretended to be our fathers because they would bring us food – like fathers. We used to receive food all the time – they would bring food to our room, but a crew member would only bring food to a person he wanted to give it to. We used to eat a lot. Ikidluak, who looks like a white person, had one of his fathers from the crew. He was the child of those we were travelling with. My own brother Peter was born on our last trip on the *Active*. It was said that his father was a white man, but that is only because he got his name from the pretending father he had. That's where he got the name Peter. His pretending father's name was Peter.

"We had a lot of the white man's canned food, but there were times when we had some of our own country food. And also, sometimes walrus meat was boiled for us. The cook was a white man. Everybody ate.

"There was a gramophone on shipboard, and the lower ranks used to play the accordion. I used to play, too. If I had an accordion now, I would play it still. I learnt on the ship, but in camp I played with my own accordion. I don't know the whalers' music, but I know the old Eskimo music. Sometimes the whalers played their own songs and sometimes they played the music from here.

"When we were getting on and off the ship we received presents from the white man – flour, everything. Cloth for dresses and knives and other things for men. Pay was never mentioned. We didn't know money then. We'd get all kinds of stuff from the ship. The things my parents received they put to their own uses, and I used my material for dresses. When I was a young girl all the young Eskimo women made all their own clothing – parkas, mitts, boots. It was not very difficult to do those things when we had all our own teeth and our eyes were healthy. But now everyone is too lazy to chew the boots.

"I used to get beads from the ship, and back in camp I used to make them into beadwork for the front of my amautiq. I got some spoons and I decorated also with those. I would cut off the handles and use the bowls on the amautiq. I used to have my parka front decorated with little metal pieces. When these little metal decorations were on the amautiq and the front was heavy and weighted down with the beads, it looked really nice. We would put coins around the tail, and they made a beautiful ringing sound. We had big ones, little ones, and medium-sized. Nowadays, no one even collects pennies.

"The boat we used to ride on would land us at a lot of places. Sometimes the Inuit alone would go after the whales without the qallunaat. Sometimes the ship left, and sometimes the Inuit would be camping until the ship picked us up in the spring. Twice we spent winters in the land of big whales. The last time[3] I think I was just becoming a teenager – I have never kept track of my age; I'm poor at keeping the track.

"After the ship left that time, the Inuit alone went out to catch the big whale. There was not one qallunaaq left behind.

"I don't know if the shamans were along on the trips because my parents were not shaman people. But even though I do not think my father was a shaman, he had a little chant. He would chant away although I never paid any attention to his chanting.

"I remember once when they were catching the big whale how the women went up to the top of a hill – a white cliff under which blueberries used to grow – to watch their husbands and our fathers catching the big whale. I was watching too. They used the spear guns, the kind the whalers brought. The gun was in the front of the boat. For a long time after the whale was hit, the arvik would not surface. It stayed down for a long time. Then many boats would tow the whale to shore. They used to be all excited when they were catching the big whale.

"They would pull the whale to a smooth beach where there was never a tide. When the whale was half in the water and half on shore, they'd start skinning. They'd make steps on the side – that was the only easy way to get to the top – and start taking the maktaaq off. They'd take the blubber and cut it into small pieces and dump it into the storage. I suppose there was a place for the fat. They also took the suka – the baleen. That's where the whale used to catch his food. The Inuit used to find it very good for getting the little fish out from under the rocks.

"The only part of the big whale we ever ate was the maktaaq. It was very much softer and better than the white whale maktaaq.

"I haven't had a piece of the arvik maktaaq since I was a little girl."

When the Whalers Were Up North

Wintering

Repulse Bay today is remote, even with modern transport, as it was in the whalers' time. Its small community has grown around the Hudson's Bay Company trading post on the shores of a protected bay into which pods of white whales venture. When belugas are sighted, all hands (except the hands in the school) lay down tools and rush to the shore to watch the hunters in their outboard canoes. Their harpoons have floats attached, sometimes the sealskin avataq or sometimes a large modern water-filled plastic container put to an age-old use.

To find evidence of the historic past you have to go out of town, so one afternoon, with interpreter Steve Kopak, I walked over the tundra about a mile and a half to a small sheltered cove. Steve had brought me here to show me the sod houses where in summertime Inuit families who worked for Captain John Murray, whom the Inuit called Nakungajuq – Cross Eyes – used to live when his vessel was whaling in the area.

Not far from shore Steve and I easily found the depressions where the sod houses once had been. We counted at least twelve, and perhaps there were more. There were some that suggested almost apartment-like complexes with several rooms and storage areas. The ground all around was springy and soft, perfect, it seemed, for sod houses, and later I learned that these houses with sod walls and canvas roofs had been warm and very comfortable. "This is where Cross Eyes' people lived," Steve said. "Captain Comer's people were a few bays farther down. The chief there, Harry, lived in a real wooden house with down mattresses."

In summer when the Scottish vessel was in sight out on the water and "something was happening, or there was news to be given out," the captain would run up a flag "so the crews would go to the ship,"

Skins pegged up to dry against the insulating snow wall banking the Era, *Cape Fullerton, George Comer, 25 March 1901.*
Mystic Seaport Museum, Mystic, Conn.

Wintering   95

Leah Arnaujaq, Repulse Bay's great storyteller, related.

Whales still swim in the water where Repulse Bay people hunt. "They never go after whales today, but you can see whales practically every day at Gore Bay," Arnaujaq told me.

These plentiful whales at the head of the sound and near Gore Bay and Lyon Inlet, where more intrepid whalers ventured, brought whalemen to the region throughout the whaling years and caused Ships Harbour Islands in Repulse Bay to be used with some frequency as an anchorage throughout the era.

After 1896 Captain Comer's favoured harbour was Cape Fullerton (Qatiktalik in Inuktitut), lower down the sound; his presence attracted Canadian government expeditions and led to the establishment there in 1903 of the first North West Mounted Police (later Royal Canadian Mounted Police) post in the eastern Arctic. But in the first years of the new century the Scottish Kinnes interests took the head of the sound and nearby waters as their base of operations, maintaining and staffing the ketch *Ernest William* as a year-round station and collecting the year's catch each summer by steam whaler. The ketch's captain for a time was John Murray, who also wintered with the *Albert* in 1912–13. He and his brother had forged their alliances in the area in the 1890s, when both had captained the *Perseverance* during a short sortie by the Hudson's Bay Company into bowhead whaling.

Leah Arnaujaq lived in the snow village built on the ice in the Ships Harbour Islands when Captain John Murray wintered over in 1912–13. She believes she is a daughter of Alexander Murray. "I knew Cross Eyes better, but Alexander was my father." Well into her eighties, Leah Arnaujaq is a much-venerated elder of the Keewatin. "She's our grandmother, she's everyone's grandmother," says Tagak Curley. Not long ago, when an official delegation of the territorial government visited Repulse Bay, Leah Arnaujaq entertained everyone back at her bungalow. And, after we did our interviews together, she invited me to the little hut behind her house, where, seated on caribou skins, I watched as Arnaujaq lit up her qulliq – the seal-oil lamp. The last time she had done so was for visiting nuns. Most quliit today are in museums. Arnaujaq showed me how to spread the flame with a little stick along the moss wick in the soapstone oval bowl that had once been the hearth and source of heat in her igloo homes of the past.

Leah Arnaujaq speaks with a vigour that belies her age and tells a story with dramatic flourish. "I remember clearly because it happened in my childhood at a time when everything is clear that one year my family travelled down by whaleboat to a place near Cape Fullerton and we waited by the river for Cross Eyes and his ship. At the time we left, Captain Comer and his ship were already in Repulse Bay. His people

This impressive igloo was built by Harry, Comer's Inuit mate. It measured 27 1/2 feet in diameter and 12 1/2 feet from the centre of the igloo to the ground. A.P. Low, 1903–04.
National Archives of Canada, c24522.

were living in cabins with down mattresses while the ship was roaming around. It was Captain Comer and his people who mostly spent their time around Whale Point, Cape Fullerton, and Coral Harbour. His ship would stay put, but the mates and the crews in the whaleboats would be all over the place. There'd be nothing but sails all around.

"When Cross Eyes came by with the ship Pauti – 'sooty' – we saw the smoke from the funnel appear on the horizon – he asked us to come on board and we travelled back up to Repulse Bay. Later on that summer Captain Comer left, but Cross Eyes and his people went down to the Harbour Islands and spent their winter there.

"Captain Comer's group and John Murray's group were rivals, so they didn't get together much. There were no battles, but now we realize looking back that they weren't too friendly – they didn't mingle. The captains didn't want to lose their Inuit people to the other ship. The Inuit working for the whalers were always friendly among themselves – most of them were relatives. People didn't think 'Those are your enemies,' but there was a line drawn. There were the people who belonged to our group and those who belonged to the other group – chiefly because the captains wanted to keep certain Inuit people for themselves.

"The Inuit and the white people used to have a lot of fun, and the relationships in those days were very close. They would go around together in good friendship.

"When winter came and the ice was on the ground, the whalers would make a big shelter on top of the deck to cover the ship. They'd make it out of wood, and when they finished building, they called the ship the 'barge.' In Inuktitut it is 'Sikauq.' The crew would eat in the shelter; the higher ranks, the captain and assistants, would eat in their cabins down in the ship. Only the Inuit leaders would be invited down to watch their entertainments. Of course, as was their custom, the qallunaat ate at regular hours – they had 'mealtimes' – so even the Inuit became controlled by time and started having breakfast, lunch, and supper.

"Everybody ate – even the children. As soon as the ice was strong and the ship was frozen in the inlet, the Inuit made their igloos round the ship so as to be near those they were working for. When the house was on top of the ship, the qallunaat used to get all the children together and count us one by one, just to make sure we were all on board. Then they'd serve us dinner. By that time we had clocks and we used to use the clock, but often the cook would get quite angry when some of the children were late. He'd complain a lot. He'd say, 'Late again!' They wanted everyone to be on time so they could count us as we boarded the ship.

"This cook who used to scold took good care of me. He'd let me do errands. He'd say, 'Go and throw the water out!' or 'Go and get some water!' I was the only child he'd tell. In those days we didn't know of flour or bread or bannock, and when I did these little errands, he'd give me some sort of pastry.

"We used to have delicious proper meals. Different kinds of porridges that were cooked in the mornings, biscuits, and bread. Everyone had coffee, even the little children. At that time I didn't know tea; I didn't know granulated sugar. I only knew about coffee and molasses. We put molasses in the coffee and it tasted rather good. They had huge coffee containers made from metal.

"The qallunaat wintering here needed outer garments, so the women made them for them. They had no measuring tapes, so the Inuit women measured with their hands. They'd use their hands on a person to measure how many hands long he was. When the garment was finished, it would fit that particular person exactly. The style was the same for qallunaaq or Inuk, but the sizes for the qallunaat were larger. Caribou for winter and sealskins for summer. I remember they used to make the whalers sealskin jackets.

"The whalers wanted practically any kinds of skin. They'd trade for musk-ox and fox pelts. Sometimes they'd get musk-ox skins even from Pelly Bay people. Do you know siksiit [Arctic ground squirrels]? One time a Pelly Bay woman who had gotten some siksiit came on to the barge and approached the cook I have told about. She said, 'I have these siksiit here in my hands and in return I want a thimble.' The cook didn't understand, so she said it again. 'Take these siksiit and give me a thimble.' He still didn't understand, but someone interpreted and he went downstairs and came back with a thimble. At that time the thimbles didn't even have tops. The lady who wanted to be paid for her siksiit got herself a thimble and left the siksiit behind. She must have been really crazy to trade all the siksiit for a thimble. The skins made a big beautiful blanket – the pelts were all sewn together into a blanket with the tails laid out and sewn around the edges.

"During the time when the Inuit had their igloos round the ship the hunters and their wives too would go down to the floe edge. In those days even the ladies went out hunting. I went too. My children would never believe me when I said that here in the Repulse Bay area when we were down at the floe edge for seal or walrus we used to see caribou crossing the ice to Beach Point on the other side of the islands. Although they went for sea mammals, they could end up catching caribou. They'd send one boat in one direction and one in the other and that way they'd end up having a variety of meat.

"As soon as the ice was out, the crews started climbing up the mast to the highest point – only those not scared of heights! – to look through the binoculars and watch for whales. Everyone would be busy with different activities. Some of the crew would automatically get in the boats. They'd have two or three whaleboats going in different directions.

"When a whale was sighted, we'd hear the cry 'ABLOW!' – there she blows! Whoever was up on the mast – Inuk or qallunaaq – would yell out 'ABLOW!' to let people know a whale was there.

"Right away they'd jump in their boats. They had no motors as we do now, so if the sea was calm they'd

When the Whalers Were Up North

"A woman chewing a boot bottom, getting it ready to sew on to the boot top," says a note appended to this photograph. Sealskin used for boots had to be chewed soft and its fibre broken. George Comer, 1907–09.

American Museum of Natural History, New York.

Preparing skins and sinew. Pitseolak Ashoona, Cape Dorset. Felt-pen drawing, 1970.

row; if there was a wind they'd use their sails and head out to catch the whale they had sighted. They'd be looking all over the place, but particularly among the Harbour Islands because there are many channels there; often that's where they'd harpoon a whale.

"When a whale was killed they'd get all the boats together and make a tow line to tow the whale to the closest shore.

"When they caught a whale the old people on the shore would tie their feet together at the ankles and play games; all the children would be tied in pairs – they'd push each other, and if they fell over they'd try to get up before someone pushed them down again. These are some of the games they'd play. And a little boy whose father was one of those who had rowed the boat – those men would get really thirsty – would carry down a bucket of water so the men would have drink to quench their thirst. It could take a long time to tow the whale if the catch was far from shore.

"When they caught a big bowhead they made a stepladder out of the whale itself so they could go up the steps and get to the top to cut it up. They had real big long-handled curved knives, and they'd start from the top and work down to the bottom. They'd take off the blubber first, and then they'd start working on the meat. The older men among the crew would take the blubber and cook it, and the meat and the maktaaq would be divided among the Inuit people. So that people would not just grab for it, one person would share it out equally, according to how many persons were in a family. The larger the family, the larger the amount. Smaller families got less.

"During the time they were cutting up the whale the men used to get really dirty and really oily, so the women would be squeamish and not want their men to sit on the sleeping platforms because they were such a sight. You didn't want them sitting near to you on your clean bed. They had special clothing for cutting up the whale. After all their duties were done and everything was put away and cleaned up, they would come home and strip their clothes. They'd be covered with oil and blood. They probably didn't smell too good. Probably their wives would help clean them up.

"The more valuable stuff – the fat – was taken by the whalers. The crew – the older people among them – used to cook the blubber, and they'd preserve the oil they cooked up in those big barrels.

"Here in this house I have a piece of whalebone from a whale my Inuit father caught over at Maliksitaq before I was able to remember. His name was Tukturjuk; his whaling name was John Bull. My grandson Simeonie brought it to me a couple of years ago. Until recently at Maliksitaq there were still some ribs – pure white – from that whale. Now people have taken them for carving, and probably only the head is left.

"My father caught that big whale on a whaling expedition for John Murray, and for getting the whale he got a boat. Whoever caught a whale got a boat. Automatically the boat he used was given to him. They would say, 'We're giving you this in thanks.' The captain would give a whaleboat to anyone who caught a whale. But Captain Comer's boats were a lot smaller-sized than Captain Murray's – and easier to handle. But you couldn't go out and borrow a boat from Captain Comer if you were a crew for Captain Murray. There was no choice. But I don't think anybody complained in those days.

"So I grew up with wooden boats. It was only after I married and my first husband and I went up to Igloolik that I saw the old Inuit umiat – the big skin boats. I kept seeing these funny-looking things made of thick walrus hide. I didn't know they were boats. When people said they were getting ready for a walrus hunt, I looked at those walrus-skin boats – just a couple of pieces of wood and walrus skins – and said, "No!" They didn't look sturdy. But the men went out, and my husband said when the umiaq got wet, the hide sank way down under his feet. He sank until the skin stopped stretching. Later we were camping where there were lots of fish, and the women were asked to go out in the umiaq and keep the fish at the mouth of the river. They asked me to go in the boat to help. I said 'Never!' but eventually I went. Funnily enough, I didn't get that sinking feeling because there was so much excitement – we were throwing rocks in the water – and later I even helped make an umiaq.

"But that sinking feeling is scary."

One always wanted to return to see Leah Arnaujaq again, and during my stay in Repulse Bay I visited three or four times. One afternoon she told me about her grandfather's trip to Scotland.

During Arctic whaling, captains sometimes took Inuit back to their homelands with them. The first Inuk voyager to Scotland was probably Eenoolooapik of Baffin Island, who in 1840 led the famous Scottish captain William Penny to the great pods of whales in Cumberland Sound. A number of Inuit, including Johnnibo, Annie Kimilu, and little Kudlarjuk, visited New England, and Inuit children even went to school there. However, there were hazards to these journeys, and journeys to Scotland at times were discouraged by British authorities. Some Inuit, too, believed it was unwise to make these voyages; on many occasions travellers did not return, succumbing, it now appears, to illnesses to which they had no immunity. A few years after the turn of the century, when the Tay Whale Fishing Company under direction of the industrious Robert Kinnes began mining mica near Lake Harbour, the Dundee vessel *Active* took Kingwatsiak, whose descendants are well known today in Cape Dorset, back to her home port. Kingwatsiak was a mica worker at the mine the vessel's owners had opened up, and the owners wanted to teach him more about the art of cutting mica. But according to Ikidluak of Lake Harbour, Kingwatsiak wasn't the one originally invited. He says, "It was Qayuarjuk, his younger brother, who was asked to go along. But his relatives didn't want Qayuarjuk to go, so when people were expecting the ship to come in, his relatives took him caribou hunting, and Kingwatsiak went instead – because he was there." Perhaps the fears were justified because Barrie Kinnes, who runs the Kinnes operations today, recalls hearing stories that Kingwatsiak came down with "every ailment known to man."[1]

Arnaujaq's grandfather, Siaraq, had to be a brave man to make the journey. "Did you ever hear or read of an Inuk crossing the ocean to the Europeans? One time whalers wanted to take an Inuk back on their journey to Scotland and it ended up being my grandfather. They had to find someone with the courage to go. He was the first to go.

"They crossed the ocean and the landscape came in sight. My grandfather wondered when they would be going ashore, and the whalers told him, 'Still further ahead,' and 'Watch for a lighthouse.' The light came in sight, and they followed a marked route that indicated where the shallows were. When they landed, the captain approached an official and told him, "We have an Inuk along." The official said, "Let him stay on board." Most of the crew had been told to leave and everyone was climbing ashore, but my grandfather had to be left behind. The captain told him to clean up the ship. If he finished the job, they'd give him all kinds of gifts. He got halfway through and then he said he was sick. The qallunaat knew he was pretending but they told him to get up on deck, and he left the ship.

"My grandfather stayed for a whole year before the ship took him back home to his country. He used to tell many stories which I heard from his adopted daughter. He said there were many things to see, and he made a song out of what he had seen:

Here I am in the land of the white man
Ai, ai, ai
I have been taken to their land,
Ai, ai, ai
Because there are so many things,
so many things to see
Ai, ai, ai
I have not noticed the days go by.
Ai, ai, ai
I remember policemen walking
with skinny legs compared to me.
Ai, ai, ai
Another day and the land is smoking
Ai, ai, ai
Because their dog teams smoke across the land.
Ai, ai, ai"

The whalers stayed in Hudson Bay for sixty-five years. When they left, around the outbreak of the First World War, there was some puzzlement, and Leah Arnaujaq thinks some people missed them.

"When the whalers left and didn't come back," she says, "I was a bit too young to feel really sad, but I think there were people who did. Every time the whalers went they would leave food behind for the Inuit people. They would give a lot of sweets and candies to the children and put biscuits inside a big barrel so they would keep. And that orange marmalade, and butter too, all in huge containers. And pork and beef with fat inside the beef. All in metal containers. They used to leave all this behind when

Umiak. Jamasie Teevie, Cape Dorset. Engraving, 1973.

they were going away. I used to think they must have so much food in order to leave so much behind.

"Some of the people may have wondered why they didn't return to Repulse Bay. After I was married and became an adult my husband and I wondered about it. We wondered if it had anything to do with the Hudson's Bay post. That was the time when the company was beginning to come into the North. We wondered, but we never really knew why the whalers didn't come back. We were kind of regretful because we remembered how good their food had tasted and we remembered everybody getting together like a big family. When the whalers left, that big family feeling was gone."

When the Whalers Were Up North

Personalities of the Bay

When I met Joe Curley he was living in Eskimo Point, the site of the Inuit Cultural Institute, set up to preserve the Inuit heritage. But he grew up in the Repulse Bay area and was born "not in the 1900s but in the late 1800s" near Lyon Inlet at a camp called Maliksitaq – "the place where the ghosts chase women." He got his qallunaaq name, he told me, because of his curly hair, and it is passed on to many of (at last count) 117 grandchildren, descendants of 24 children by 2 wives.

Aivilik women. Because the photographer Geraldine Moodie also photographed the women in this picture individually and noted their Inuit names, it is possible to give names to all the people in this photograph. Most are still remembered on the west coast of Hudson Bay, although southern pronunciation and phonetic spelling sometimes complicate identification. From left to right: Kookooleshook (Kukilasak), Tuucklucklock, Nivisinark (Shoofly), Towtook, Kooalashalolay, Uckonuck, 1904–05.

Museum of Mankind, British Museum, London.

At the time of Joe Curley's birth, the whaling era was entering its last phase. Furs were fast becoming as important as whale-oil and baleen. And in 1903, with the coming of the Royal North West Mounted Police to Hudson Bay, whalers and Inuit began to face regulation. The force collected taxes, looked askance at musk-ox hunting, and began its famous long patrols. Life in the North was changing.

As a child of a whaling boss Joe Curley saw the final whaling days from a privileged vantage point. He knew all the personalities of the day: John Murray, Alexander Murray, George Comer, and George W. Cleveland, the man the Inuit called Suquortaronik – the harpooner. When the whaling days were done, the famous Shoofly, Comer's close companion, became his stepmother. During our talks Joe Curley made these personalities, all long dead, seem vivid and alive. Much of what I learned about them came from him, although he sometimes directed me to others who he thought would have more to tell.

I. SCOTCH TOM

One of the dominant figures of the time was Joe Curley's own uncle and adopting father, Angutimmarik, or Scotch Tom. Whaling masters had adopted the practice of having the Inuit boss hire and direct the Inuit crews. The system outlasted the whalers' day and was inherited and continued by the fox-fur traders who followed the whalers. The chief native was known as "the boss," "the mate," or "the first trade man, because he was the first to trade." For many years Scotch Tom was John Murray's Inuit mate. His opposite number was Harry Ippaktuq – "dirt" – Tasseok, who whaled for Comer. In this brief memoir of his childhood among the whalers Joe Curley tells us something of his father's role.

"My father was the captain's man. In those days the Scotch and the Americans had their Inuit leaders, and my father Scotch Tom, my uncle who adopted me – his Inuit name was Angutimmarik – was the man for the Scottish people. That was the tradition in the past. The whalers would pick a person as chief, and he would become the most powerful person. When Angutimmarik was a very young man, when the Scots first started visiting the different camps up here, they took him to work for them because he was a good hunter and a reliable man.

"The Captain's name was John Murray. We called him Nakungajuq – Cross Eyes. I remember the year, early in the century, that his ship came into Repulse Bay. It is logged on the rock in the Harbour Islands.

Joe Curley, also called Kayakjuak after his grandfather. Southampton Island, May 1926.
L.T. Burwash. National Archives of Canada, PA99404.

That year there were two ships docked. There were two Murray brothers, Alexander and John, and they both had their own ships; they were Scottish ships. On one of the Harbour Islands the whalers had made a landmark on a flat rock. They used to write there and indicate the year they arrived. When I was a child I watched them chisel away on the stone and carve a bowhead whale. Cross Eyes stayed three years – if

The Scotch ketch with whale in tow. Joe Curley, Eskimo Point. Coloured-pencil and biro drawing, 1983.

they had a good supply of food, they'd winter as long as they could. Those brothers each had their own ship, although I don't know that they travelled together much.

"There were whalers all around when I started realizing and remembering things. My original parents, Mike Kanajuq and Angearok his wife, came over from Pond Inlet. They travelled by dog-team over to Igloolik, and I was born at Maliksitaq camp near Lyon Inlet, over past Repulse Bay, in the late 1800s. All my parents, my original parents and Angutimmarik and Agglak who adopted me, were working for the whalers. At that time there would be six whaleboats with the ship, and in each there would be five Inuit people. The Inuit leader would choose the people. We always had relatives working for the whalers, and when we were working for the whalers, we were always on the move. Relatives could not always be together. The relatives we had were scattered all over the place.

"After whaling ended, Mike Kanajuq went back to Pangnirtung, and I've relations there today. If you see anybody there with curly hair, you'll know he's related to me.

"The Scots and the Americans made their arrangements through the Inuit chiefs. My father had his own whaleboat, which he'd been given with all the equipment inside so he could catch whales for them. He would be giving out orders to the Inuit helping the whalers and also to his own men. They'd hunt whales and walrus. In April and May, at the start of the whaling season, they'd get prepared and go down to where the whales might be. This would be at the floe edge, near where they were camping. There'd be six whaleboats with Inuit and qallunaat whalers. Sometimes they'd go out together, sometimes in separate boats. They would sing when they were towing a whale ashore – while we on the shore were dancing!

"It was usually the white people who cut up the whale and put the blubber in the barrels. The whalers would tow the whale up on the shore, helpers would do the rest of the work, and the whalers would go back down to the sea again.

"Sometimes we'd set up our tents and camp on the shore; sometimes we'd go out on the whaling vessel. If the whalers were going any distance, they'd always take the Inuit along. We would spend a long month on board.

"As a child I had to wear a harness. They'd tie me anywhere on the ship. They started tying me up because I used to run around the deck and one time I fell off. The women were sleeping on the deck in a corner at the front of the ship. It was because I didn't want to trample over them and had just avoided falling on one of them that I tumbled into the water. It was the middle of nowhere; no land was around. I shouted, but it was quite some time before they noticed me and came to pick me up. The white people had been teaching me to swim. In the summer, although some of the lakes are very very deep, there are some that are shallow and fairly warm, so we used to swim in these waters which were not too deep. So with strokes, sort of strides, I was able to float on top

When the Whalers Were Up North

of the water and swim towards them.

"That's how bad I was. But they didn't even scold me.

"I started travelling with them as a young child and continued until I was fourteen or fifteen. At that time the second mate was teaching me English, but he died before I could learn much. He died of mumps, and his grave is down in the Harbour Islands.

"It was pretty much the same, working with Americans or Scots. The men would be out for the whole summer and finally come back at the end of the whaling season to their families and friends. It was all the same, Americans or Scots, their lifestyles were much the same. The American whalers who came in had a lot of hospitality towards the Inuit people. The Scottish people did not have the possessions the Americans had, so they appeared to give less. But John Murray sent my father lots of gifts and parcels. My father must have been working for the whalers long before I was born. You could tell that he must have been because of his possessions. He had a lot of weapons, all kinds of rifles. He was always receiving gifts, and many were expensive."

II. CAPTAIN COMER, THE WHITE SHAMAN

The missionaries established their missions in Hudson Bay and the strait[1] as the whaling era drew to a close, and sometimes today Inuit prefer not to talk about the shamans. But wherever I visited, up and down the Keewatin coast, people liked to talk about Captain Comer, the white shaman, and to look at the magical images he made appear, some of which showed his opposite numbers, the Inuit shamans, at work. "Ah, Angakkuq with his little bald head; he was a shamanist parson!" murmured Leah Arnaujaq of Repulse Bay with a touch of nostalgia as she looked at some of his photographs. And Kanajuq Bruce, of Coral Harbour, great-great-niece of Harry, Comer's Inuit mate, remarked, "Angakkuq was a very likeable man. He got his name because of the photographs and because he had those little technical things that would wind up. People here never used to have those mechanical things. Every so often Harry was invited to dinner with the captain, and then in turn Angakkuq would share with the Inuit and eat some seal meat."

Joe Curley filled out the picture: "The Americans were mostly around Qatiktalik – Cape Fullerton. They would winter there and then sail up towards Repulse Bay. We didn't travel with the Americans – they had their own Inuit, Harry and his group – but I remember Captain Comer. We used to call him Angakkuq – the shaman – because he was able to take photographs. They would appear just like that, out of a piece of paper.

"Quite often in the summertime we used to meet them. You often heard people talking about them, so we knew who they were. We would see them during the summer months around Cape Fullerton and around Marble Island and Chesterfield Inlet and Rankin Inlet. They would be hunting for whales – beluga whales, bowhead whales – and walrus and polar-bear hides.

"Once my father was given an assignment to bring goods to the place the Americans were going to make their destination. That's the only time we were with the Americans – when we took their goods to them. The Scottish and the Americans were on good terms, but they didn't really get together.

"We used to hear a lot about the captain taking photographs. He had a camera he would cover up with a cloth. He had glass negatives. But I don't remember if he had Inuit helping develop. He brought in different pieces of equipment. That's how he started taking photographs of the Inuit.

"People thought he was a shaman. An Inuk knows a shaman as a person who is able to make things appear or happen – in a way no ordinary person can. If he had not produced these magical things – so exciting and so fascinating – they would not have called him Angakkuq. But that is exactly what they called him – Angakkuq – a shaman. He didn't mind. Angakkuq was his nickname. But people didn't respect him so highly as a shaman that they weren't able to talk to him. They could have a normal conversation.

"Did the photographs scare people? I don't remember the Inuit being scared of him. They made friends with him and he used to eat among them. They favoured him a lot. They called him the Angakkuq because he was able to do wonders."

Comer began to use a camera in the Arctic in 1893; a fellow whaling captain is said to have given him the first camera. Over his years in the North he took hundreds of pictures, compiling a magnificent record of the whaling days and of Inuit life at the time. The task was not easy. In the *Era* journal for 1 February 1904 Comer wrote of the difficulty he experienced in photographing Inuit in the *Era* cabin: "The first flash light I probably did not fix right as it came near taking the roof of the cabin off – at least the report sounded loud enough to have done so."[2] And if Inuit were fascinated by Comer's special brand of magic, Angakkuq, as he is still recalled up and down the west coast of Hudson Bay, returned the compliment. He acquired remarkable photographs of his Inuit mate Harry demonstrating his magical activities. And in his journals are many references to the Inuit shamans and to their seances, which he called "anticoots."

"Last night the natives held a meeting," Comer wrote on 14 September 1897, "and one of them who pretends to have power to visit the spirit land went into a trance to call upon the spirits who have taken charge of the whales and learn why we had not had better success. This morning the natives told me that the anticok was told by the deities who look after the whales that we had worked on the musk-ox skins at a season when we had not ought to, or perhaps we worked on them on the vessel and that being on the water was wrong. Also we had an owl skin which we had saved and we had done wrong again in picking the feathers off the ducks. We had done wrong, we should have skinned them. I was told to throw up my hands several times as though throwing things away. After I had done that (which I did) they told

Shamans demonstrated their practices for Comer's camera.
Though tightly bound this man will escape from his ropes.
George Comer. 1900–05.

Mystic Seaport Museum, Mystic, Conn.

me I was alright now and would have good success in the future."[3]

Joe Curley sometimes said he thought that, besides his photographic activities, there might be another reason why Angakkuq received the nickname he did: he asked so many questions about the shamans who were often helpful to the whalers. "Captain Comer was deeply interested in shamanism," Joe Curley remarked. "So he could be a powerful captain, he relied

a great deal on the shamans. He needed their help during the whaling expeditions. That's why he kept company with those among the Inuit who were shamans."

Comer set out to record Inuit traditions. He worked closely with Franz Boas, the foremost anthropologist of his time, and passed on insights that broadened knowledge of Inuit life. "Reading and writing is the way I am passing the time," he wrote in his diary, frozen in on 11 November 1903, a snowy, stormy day. "I endeavour to write down all native customs and stories."[4] He had also taken with him a graphophone and fifty blanks with which to make records of songs and speech. Another project involved the making of plaster casts of Inuit faces. Because of the rapid cultural change in many traditional societies at the time, Boas, who commissioned Comer to bring him artifacts, was a proponent of the practice, common among anthropologists then, of collecting plaster casts to preserve a record of distinctive physical characteristics. A few years earlier he had arranged for Comer to have lessons in making the casts. "If you have time now, I should like to send Mr. Mayer, our sculptor, to East Haddam to give you some more instruction in the taking of plaster casts," wrote Boas to Comer on 9 May, before he left on his 1900 voyage to the Bay.[5] Shortly thereafter, three barrels of plaster were delivered to the *Era*, and Comer, whose own plaster bust was taken by Mayer, eventually made some hundreds of casts, all now in the American Museum of Natural History.

"He used to mix up a great paste and make sculpture out of the paste," said Joe Curley. "Inuit people used to say he was putting too much water into his mixture. He was always at work, doing things, probably as hobbies, on his own time. I don't know what he did with the masks he made of people. He took them down south to the States and probably had them made up properly in the States. These sculptures were chiefly of Harry's group – Harry and his workers. All of them had their masks made by the captain. He may have made them a little tight on the faces. Getting them off, that may have been a problem, but I never saw him at work. It was well known that Captain Comer was always busy."

III. SHOOFLY

With the passage of time Inuit families have forgotten some of the whaling names of their parents and grandparents. But no one has forgotten that Shoofly was the whaling name of Nivisinaaq, Captain Comer's close companion and, perhaps because so many qallunaat took her picture, the most famous of all the Inuit women whose relations with qallunaat men could be both warm and enduring.

The name Shoofly comes from an old song popular in Civil War times – *Shew Fly, Don't Bother Me* – and first written down in 1869. (*Shoo-Fly Pie and Apple Pan Dowdy*, which Dinah Shore made famous, was composed in 1945.)[6] "Her captain gave her that name because she was always shooing away the flies," says

a grandson on Southampton Island, where, like everywhere in the Arctic, mosquitoes come in plagues.[7]

Shoofly died in the 1930s, tired and sick, but Joe Curley knew her in her prime. With her son John Ell (after John L. Sullivan) she travelled with Captain Comer on board ship. "I can't tell you for how many years Shoofly and Angakkuq were living together, but they were always together. Whenever he came up here she would travel with the captain. They were looked upon as man and wife.

"When the whalers left, Shoofly ended up living with my father Scotch Tom. Shoofly came to live with us and became my stepmother. She was a wise person, and well organized in most of the things she did. People looked to her for advice, and she was really good at dealing with people. I'll take myself as an example: I have a very quick temper, and she was able to tell me she realized the reason behind it. If she knew anyone was going in the wrong direction in life, she could advise them in a few words. She was always looking out for other people. Even though she was a woman, she was regarded as one of the leaders among our people. She was what you would call today a superior person."

One felt the need to know more about Shoofly, and I sought out her granddaughter Bernadette Ookpik Patterk, who lives in Rankin Inlet with her husband Joe and their family. Joe's ancestors, too, were whalers; his father worked for John Murray, among other things keeping a check, once the whalers had left for their home ports, on the wooden shack the

Scots built at Whale Point. Joe has a whaling name he inherited – "Coopie." It comes from "coffee." Joe says, "Maybe every morning the whalers would yell out "Coofie, Coofie!"

Born about 1917, Bernadette Ookpik – she gets her first name from the French missionaries – is the daughter of Shoofly's only biological son, John Ell, whose Inuit name was Oudlynnock. John Ell participated in many Arctic expeditions, and has been called the best-known Eskimo of his generation. He read and wrote English, played chess ("the game with the little pieces that look like horses' heads," a youthful interpreter explained), and took and developed photographs. Inuit often refer to him as Comer's stepson; some declare he was Comer's actual son. Ookpik does not consider this likely.

"Yes, my grandmother had a captain friend," she said. "In fact, she had pictures of the captain and used to receive presents of clothes and cheques. But it was after John Ell was born that Angakkuq was living with Shoofly. My grandmother could bear no more children. She had to adopt them. John Ell had no qallunaaq blood. He was true Inuit."

Shoofly was one of the wives of Tugaak, known as Ben, but liaisons between Inuit women and the qallunaat were commonplace during the whaling years and accepted. "Ben knew the captain had Nivisinaaq as a girlfriend, but they were real good friends to each other," Ookpik said. When Ben died, Comer recorded in his diary that he helped with the burial.[8]

In the winter of 1893–94 Ben had, in fact, saved Comer's life. "This is the man who pulled me out of the water when I had broken through the thin ice and to whose timely arrival I owe my life," Comer wrote as Ben lay dying.[9] Three days later he wrote, "Our native Ben who has been sick died this afternoon ... I helped carry him away and assisted in the burial. His two wives, brother Harry, and his children, also Gilbert and Sam, we built stones around and over him. Our mittens had to be thrown away. The ropes which held the deer skin around him had to be all cut so that he could rest easy."[10]

Ookpik had her own picture of her grandmother and brought it out to show me. Shoofly's appeal is evident. She has emphasized with paint the traditional tattoos for the camera and is wearing her long hair in laced braids in the manner sometimes said to indicate a woman has borne a male child.[11] She is wearing her most famous amautiq, purchased by Comer in 1906 for the American Museum of Natural History. On the front, besides the high-heeled boots and compasses, her name Shoofly, hidden in the picture by her braid, is embroidered in white beads. On the back a kneeling hunter takes aim at a caribou.[12]

The picture came from a folio of old photographs of Inuit life published by the Northwest Territorial government; the photograph is not by Comer but by A.P. Low, commander of the Canadian government expedition vessel *Neptune*, which wintered beside the *Era* at Cape Fullerton in 1903–04. That year and in the following year many photographs were taken of Shoofly. Also on board the wintering *Neptune* was

Superintendent John Douglas Moodie, charged with the task of establishing the first Royal Canadian (then, the North West) Mounted Police post in the Hudson Bay District at Cape Fullerton. For the ice-bound personnel of the wintering vessels picture taking enlivened the long winters. "Today got several of the women to tattoo their faces with paint, as the tattooing on their faces will not take and show in a photograph," Comer wrote in his journal of 16 February 1904. "In this way, I got five very good pictures showing as many different tribes – the Iwilic, the Netchilic, the Kenepetu, the Ponds Bay, and Southampton styles. Commander Low and Major Moodie also took pictures of the same. The commander came over this evening and gave me some good instructions in developing the plates."[13] The following year Moodie's wife Geraldine, a talented photographer,[14] joined him. Moodie is remembered as something of a martinet.

Shoofly. A.P. Low, 1903–04.   National Archives of Canada, 853548.

The Canadian government expedition vessel Arctic *against iceberg. Since 1884, shortly after the Arctic islands through treaty with Britain became Canadian territory, the Canadian government had sent a number of expeditions to Hudson Bay and the Arctic islands, but expeditions of 1903 and 1904 carried personnel charged with establishing a police post – "the first patch of white people" – on the west coast of Hudson Bay. Henceforth the whalers – and Inuit – would face regulation. J.D. Moodie or Geraldine Moodie, 1904–05.*   Courtesy of Joan Eldridge.

When the Whalers Were Up North

"He was inclined to be a tough officer to please," says his great-great-granddaughter Joan Eldridge of Vancouver, "but she was always very nice and rather interesting, or I should say interested in others."[15] Ookpik and Joe Patterk recall that a member of Shoofly's extended family became their helper. Both the Moodies took photographs of documentary value, but Geraldine's had particular flair. Her subjects were often Inuit women in their elaborate parkas.[16]

Ookpik told me that some parka decorations that had belonged to Shoofly once came into her possession. "My grandmother gave her amautiq decorations to her daughter-in-law. My mother had them, and later they were given to me. But they got lost. We were moving to the mainland from Southampton Island, and we had to leave much of our stuff behind because we had only a small boat. I never saw them again. I heard they were given to an older person. All I can remember of those decorations were the bands for the sleeves.

"The beads came from the whalers, and it was the desire of every young girl to have an amautiq like that."

Ookpik recalls her grandmother as an extremely practical person who knew her materials. "When the ship wintered, Shoofly used to help with the cleaning. At that time they were bartering services, and Shoofly was in charge – she would organize the trading. There was a young girl, Maani, and she was looking over what Shoofly was handing out – there were materials for dresses and some were very pretty; some were plain but thicker. Maani was given the thick material instead of one of the pretty ones and she didn't like that. She preferred the pretty ones, which were thin. But later she knew that Shoofly had given her the best material – material that would last. Shoofly said, 'I want you to have some wool.' "

Though Comer's relationship with Shoofly was in the tradition of the country and a rather usual one, it upset a number of members of the Canadian government expedition, the doctor included. "A very bad influence on the members of the crew," Lorris Elijah Bordon later wrote in *Memoirs of a Pioneer Doctor*.[17] It can be presumed that Inspector Moodie frowned on the relationship, too. Besides his wife, Moodie's son A.W. Moodie was also along as a special constable. Whether at this time or in later years, young Moodie took a wife from the North. Thereafter, his father had nothing to do with him and would not receive his children.[18]

Comer's journal shows his concern for Shoofly during an illness she had in the fall of 1904, when the Canadian government vessel *Arctic*, which had taken the place of the *Neptune*, was wintering beside the *Era* at Cape Fullerton. On 4 November he writes, "Two families of our natives came back today. One of the women – Shoofly – has a heavy cold on her lungs with quite a fever, had to be helped off the sled and on board the vessel. The doctor from the steamer *Arctic* has been over twice and is now taking care of her." On 6 November, Shoofly's condition had improved. "The sick woman is slightly better, though still in a dangerous condition. The doctor comes each morning and evening and only allows her to take malted milk. We still have a few oranges left and those taste well to her."[19]

Many pictures of Shoofly appear among the captain's photographs, some identified, some not. One appears to show Shoofly, her husband Ben, her "husband-sharing woman" Melia, and two children of the family. In Comer's day most Inuit men had two wives, and people frequently say what good friends the wives in a family were. "My grandmother used to joke with her husband-sharing partner about how, after they died, they must have their boxes close together. They had been great friends in life and wanted to stay together." Another told of the additional heartbreak that followed the death of the man in the family. "My grandmother cried when my grandfather's other wife was going to another family; they didn't want to be separated."

Tragedy struck Shoofly's Inuit family when, in 1909, a few years after the establishment of the police detachment at Cape Fullerton, Melia's son, Kumanark, whose whaling name was Charlie, was accidentally shot by a member of the force. The bare bones of the story appear in official police records in a dispatch written by Constable Charles R. MacMillan:

Sir, – I have the honour to report the following painful occurrence. On the evening of June 22, about 6 p.m., Const. Walker, while looking out on the ice with a telescope, said he saw a doujug, which is a large seal. Const. McDiamid and myself also looked and thought likewise. I said I would go and try to get a shot at it. The distance appeared to be about two miles. I walked to within 600 yards or so of the object, and then lay down on the ice to crawl nearer without alarming the animal. I crawled towards it for about 200 yards or more ... At last I fired a shot, and immediately the object disappeared. On standing up I saw it, and immediately ran towards it. I ran for some yards when I suddenly stopped horrified, as I saw it was a man lying face up ... [20]

Inuit people know a fuller story· "Kumanark was a very good accordion player. The ladies wanted to dance and they danced all night. When they had finished, Kumanark was tired after playing all night, and so he went out and fell asleep on the sea ice.

"This happened shortly after the first RCMP buildings went up, and an RCMP was looking out with binoculars and saw a seal on the ice. He told someone he was going out after the seal. On the ice Kumanark kept raising his head, and it was as he raised his head that he was shot. After he had fired the shot the RCMP found he had killed Kumanark. They had been good friends. He carried him home to the settlement on his back, covered in Kumanark's blood.

"Kumanark and his wife Kukilasak were visiting from another camp, and so the RCMP went to find Kukilasak.

"The RCMP was very distraught and kept handing the rifle again and again to Kukilasak, asking her to shoot him. But Kukilasak didn't want to kill a human being.

When the Whalers Were Up North

"The people felt sympathy for the RCMP and said, 'Anyone shoot him who can.' There was no hard feeling. Because Kumanark and the RCMP had been good friends.

"He was one of the first RCMPs in the North."[21]

After our talks Ookpik suggested I might find out more about Shoofly from Eugenie Tautoonie Kablutok, a lively older person who gives talks at the Rankin Inlet school about the old way of life. As a small child she lived with Shoofly and John Ell. She remembers that Shoofly was an innovator: "All the people knew that Nivisinaaq was the first to use a sewing machine in the North. She was really good and could sew really well. She started sewing cloth material for friends and relatives – that's when we started using cloth for clothes. I'm not sure what kind of cloth exactly – I've heard she used to make dresses out of thin cotton. She used to make them on the ship. The captain must have taught her how to make a pattern for dresses, and she started making dresses and skirts. Maybe they had the cloth material on Baffin Island or in the Arctic Quebec region, but people remember that it was Nivisinaaq who started the new clothing up here.

This coloured-pencil and felt-pen drawing gives the artist's impression of the arrival of the first policeman at her family's camp. Napatchie Pootoogook, Cape Dorset. 1981.

Kukilasak (Kookooleshook) with her child by another partner. Geraldine Moodie, 1904–05.

Museum of Mankind, British Museum, London.

"That new clothing became really useful. People would keep the same dress from cotton material for about two summers, sometimes more, washing it carefully and making sure there were no tears and trying to keep it as pretty as possible. Then once winter came we'd store them away where nobody would touch them. Only in summer would we wear the cotton skirts again."

Another woman recalled that Shoofly used to trade the long cotton skirts for beadwork.[22] The bead-trimmed parkas, often works of art, were collected by Comer for southern institutions. The decoration on one amautiq in the collection of the University Museum of Archaeology and Anthropology at the University of Pennsylvania appears to show Comer's *Era* under an Arctic sky.[23]

Shoofly is remembered as a pioneer and trendsetter in another sense, too. As was the custom, she adopted a number of children. "Nivisinaaq bore only one child, so she adopted more children from other people," Eugenie Tautoonie Kaklutok explained. "Inuit adopt children because they know they will grow old and become sick. If her oldest son had happened to die, she would have had no child to care for her; so she adopted so she would have some support and help. She adopted children who were half and half. Probably she was not the only one who adopted this

Inuit woman with the whaling name of Suzy, in beaded amautiq made for Lady Grey, wife of the Canadian governor general. Geraldine Moodie, 1906. Courtesy of Joan Eldridge.

kind of child, but in our area she was the first to adopt the half-breed children.''

Inuit women believe that half-white babies were generally loved and cared for, perhaps because ''in those days so few children were being born.'' And a woman explained, ''The mother would breast-feed the baby for a long time, and during this time she would not have another child – it was birth control. The child would be with the mother all the time, and the couple would think of the child as their own.''

However, in a letter to explorer Vilhajalmur Stefansson, Comer suggested that while many children were fathered by white whalers, ''there is not the affection for them by the Native Father that he has for a full-blooded child.'' In the dreadful famines that sometimes occasioned cannibalism in order that the hunters might survive, ''children of the white man are the first to be allowed to starve or perish.''[24]

Some of Shoofly's adopted children sailed with her on board ship, and it is remembered that, after the whaling days were done, the captain continued to send up ''clothes and other little things.''

When Comer left the North, he gave his photographic equipment to John Ell – ''All the cameras and photo-finishing equipment,'' says John Ell's adopted son Ben Ell, who lives in Iqaluit. There were also many photographs. When Ben Ell looked through the pack of Comer's pictures, he said, ''I have seen them before. There were stacks and stacks of these pictures. All with writing on the back.'' All are lost today. ''John Ell moved a lot,'' says Ben.[25]

Inuit children at Cape Fullerton. Geraldine Moodie, 1904–05.
Museum of Mankind, British Museum, London.

As for Shoofly, her captain never forgot her. Until her death he continued to send her presents and, as John Patterk, a great-grandson says, ''her pension.'' Shoofly's whaling name continues. When a great-great-granddaughter was born recently, the baby received the name Shoofly.

IV. SUQUORTARONIK THE HARPOONER

''Near Igloolik,'' an interpreter told me, ''there's a beach covered with caps and liquor bottles; they say that's where Suquortaronik had his last party.''[26]

George Grover Washington Cleveland of Beetlebung Corner, Martha's Vineyard, also known as Suquortaronik – the harpooner – is a name that crops up all along the whaling route. An elderly Cape Dorset woman remarked, "I certainly know of Suquortaronik, the harpooner. He went off the ships to hunt musk-ox and left many children in the camps."[27] Suquortaronik became known for the number of his progeny, twins included.

His somewhat chequered career attracted attention in the south when Peter Freuchen – Big Pete in the Keewatin – who met him about 1922, painted him as larger than life in his stories of the Fifth Thule Expedition. He wrote of him, "Cleveland was a great character. When we asked him, during our first meal together, whether he would object to our bringing out a bottle of our famous Danish schnapps, he assured us that we could make ourselves at home in his house as long as we desired. 'In fact,' he assured us, 'liquor is my favourite drink – any kind and any brand.' "[28]

Joe Curley elaborated on his career: "Oh, I knew George. He was one of the Americans, but because he was such a thief, he was just left behind among the Inuit people. He was fired. Eventually he was adopted by Harry and his family. They were concerned for him and looked after him.

"He had stolen quite a bit of equipment, mainly perhaps liquor. He stole from the ship and the crew. So they just left him here. The Inuit looked after him, fed him, and gave him clothing. They treated him as one of themselves. They didn't want him starving.

"While he was stranded, he adopted the Inuit lifestyle. He ate among us and lived with us in the camps. In winter he stayed in the igloos and in summer he went out camping in the tents. He did everything the Inuit did. I don't know if he had his own dog-team, but he went out hunting. He learned all the techniques, all the things Inuit used to do, and finally he could keep himself alive. After a while he got taken up by the Scottish whalers. But they weren't satisfied either, and it ended up that he was given back to the Inuit people. He kept being handed back and forth.

"When the HBC started up in the North, he began working for the company. Finally he got his own building and ran a store."

Cleveland seems to have first appeared in the Bay on the bark *A.R. Tucker* in 1895 and sailed again on the schooner *Francis Allyn* in 1897 and 1899. In 1900 he was put ashore at the Wager River to run a whaling and trading station for Thomas Luce & Co., the owners of Comer's vessel, the *Era*. He built a twenty-four-by twelve-foot wooden shack and for two winters had with him there a white companion, Charles Clemmons of Torrington, Connecticut. Then, when relations with Thomas Luce & Co. were apparently severed – Comer is said to have sent the crew of the *Era* to burn his station down – Cleveland decided it "was a case of getting among the Eskimos or starve, and I made for the Iwillick tribe. I found the tribe and was taken among them as one of their own brethren.

"I was with a tribe of about 300 persons, and took part in all of their wanderings ... "[29]

He found employment again with the Scottish firm of Robert Kinnes of Dundee, which had the ketch *Ernest William* in the Repulse Bay area. After the Hudson's Bay Company arrived to pursue the fox-fur trade, he worked for many years for the company.

At one point early in Suquortaronik's career he pursued a rather bizarre sideline. Besides plaster casts, Franz Boas also desired examples of Inuit skeletons and skulls for the American Museum of Natural History. "The more you bring of these the better," Boas wrote to Comer just prior to his 1900 expedition to the Bay. But Comer went about the task reluctantly, according to Arctic historian W. Gillies Ross. He asked for and received permission from the Inuit to collect skeletons as long as he left a gift in each grave for the dead person. Eventually he collected eleven or twelve skulls from ancient graves or the unoccupied houses of the extinct Sadlimiut, but none, Ross says, from graves of the Aivilik or Qaernermiut who customarily worked for his vessel. "The small number and limited selection suggest that, despite the assurance of this ship's natives, he was reluctant to disturb their graves. Perhaps he suspected that at heart they were not pleased to have him expropriate the bones of their ancestors."[30]

Comer's instincts appear to have been right. Joan Attuat of Rankin Inlet related the story, which she heard from her mother and stepfather, of how while whaling with Uvinik (the adopted son of Harry, Comer's Inuit mate) and Tugaak (Shoofly's husband), Suquortaronik suffered a curious upset.

"I've heard that the whalers used to gather up Inuit skulls. Is there anything written about this? Has anybody heard?

"My mother Maani and her husband Anguti used to tell stories about the whalers and about the shamans, and they often told about how Suquortaronik collected Inuit bones from all over, starting from Repulse Bay. Any kinds of bones. When people died in those days they were not buried in caskets or in boxes; they just buried them with stones. That's why the skulls could be collected.

"We think the reason he knew where the bones were was because the children told him. We suspect the children were collecting the bones and skulls from the graves because they knew the white people wanted them. They had seen them packing them away.

"Suquortaronik used to gather the skulls and put them in a box. One time he had all his bones, collected from different places, in the whaleboat while he and two other white men were chasing the whale. His Inuit helpers knew the boat was carrying a load of Inuit skulls, but they didn't know why, or what the skulls were to be used for. Probably he just had them along and meant to put them away with his collection when he got to shore. But he had an accident. While he was following the whale, his boat got hit by the

whale's tail and turned over. He had been warned by Uvinik and Tugaak, expert whalers and members of his whaling crew. He really had no choice but to follow their advice because they knew more about whaling than he did. The whale was harpooned and wounded and surfacing here and there. 'You better stay away from that whale,' one of them said. 'He's going to whip you with the tail or with the flipper when he surfaces and the boat is going to capsize.' But Suquortaronik paid no attention and that is exactly what happened.

"He didn't drown because Uvinik and Tugaak were nearby in their boat and rescued him. But he lost all his bones."

Suquortaronik stayed in the Bay for more than a quarter of a century and is remembered favourably: "He could be bitchy, but really he was all right"; "On the whole people seemed to like him fairly well." He left the North for the final time in the middle 1920s, when illness forced his evacuation. Joan Attuat says, "We have a watch from him. When he was old and got sick it was decided to take him to Churchill. My husband led the dogs. They travelled by dog team to Churchill, and my husband ran in front of the dogs all day. When they arrived, Suquortaronik gave my husband a watch with chimes.

"Because he had run in front of the dogs all day. And because Suquortaronik knew he was going to die."

Inuit grave with the dead man's drinking cup and hunting equipment placed nearby, about 1923.
Notman Photographic Archives, McCord Museum of Canadian History, Montreal.

When the Whalers Were Up North

CHAPTER 11

The Last Vessel at Marble Island

Once or twice a summer Robert Tatty takes out the qallunaat in his extra-large Peterhead the *Windy Bird* to visit Marble Island, the most storied of all the winter harbours. It's a two-hour journey down Rankin Inlet, then eight miles into the Bay. Inuit call the island Uqsuriaq – "calm." The Keewatin is mirage country, and on very still days you can see the island, huge, white, mysterious in the sky, "like caribou fat rising up from the water," as an interpreter remarked.

The day I went to visit Uqsuriaq the waters in the inlet were glassy still, covered with drifting seaweed bladders. Captain Tatty's family crew brought out on deck a big upholstered couch, and many of us crowded on it, watching the landscape, drinking tea, and eating the fresh bannock Mrs Tatty had provided. Our mood was anticipatory. Terrible things, wonderful things had happened at Marble Island. As we sat on

Marble Island. Ruth Annaqtuusi, Baker Lake. Linocut and stencil, 1980.

deck we reviewed legend, history, and rumour. The island with its famous natural harbour had been the scene of shipwrecks, lonely funerals, starvations, possibly of murder. It had also seen boisterous shipboard life, dancing on the rocks, minstrel shows, banjo concerts, and theatrical productions staged in a theatre constructed out of Arctic rocks. The island achieved its fame in the years after 1860 when American whalers made it for a number of decades their favourite wintering site. But mysteries and stories go back further; in 1719 the two vessels of the Knight expedition, in search of copper and riches for the Hudson's Bay Company, were wrecked there, and over two winters the crews perished. There are rumours of caves filled with bones.

Deadman's Island with its lonely graves comes first into view. It is a rocky shoal, barely detached from Marble Island, and a vessel must pass in its shelter to reach the narrow, almost hidden passage that leads to Marble Island's natural harbour. This is the famous gut through which the whalers sailed, sometimes towed, their vessels at high tide. The *Windy Bird* moved easily over the waters and entered the basin beyond. A mile long, half a mile wide, with blazing white quartzite walls, it is a breathtaking sight. Like the highest Rockies or Niagara Falls, Marble Island cannot disappoint. It is one of the great dramatic landscapes of the continent. Here in the whaling era five or six vessels sometimes wintered, anchoring near the only break in the "marble" walls, a gravelly beach that extends into a green meadow beyond. Some distance up is the freshwater pond where the whalers got their water.

As one approaches, at the far right of the beach are the jumbled foundation rocks of the whalemen's theatre. In 1864 the *Orray Taft*'s master, George J. Parker, had a detail of twenty-five whalers consigned to building the theatre (out of casks when stones could not be pried from the frozen ground) and planning and performing shows. "Have a building 30 by 42 feet as nice a one as ever was built in Hudson Bay ... " Parker wrote in the *Taft*'s journal for 16 October 1864.[1] The season was opened by the "Hudson Bay Minstrels," and the four captains of vessels wintering at Depot Island travelled down by dog-team to see the "Artic Elephant" perform.[2]

"The older people used to say that Marble Island was an iceberg that turned to land," said Captain Tatty, his daughter-in-law Sally interpreting. "That's what they used to believe." Do the Inuit find the island scary? "Oh yes, I used to share those feelings myself! They say you must crawl ashore. If a person steps ashore without crawling, a year to the day he dies."

For the first decades of whaling, whaling captains favoured Marble Island as a harbour because in spring they could be quicker out of the ice and cruise perhaps a month earlier than from anchorages up the sound. But the vessels relied on Inuit hunters crossing from the mainland and bringing fresh food to the island. Some years, if the weather and conditions were bad, the Inuit might not come. Most of the graves on Deadman's Island are those of victims of scurvy. In the later days of whaling, Marble Island lost its popularity.

One vessel at least, however, wintered there early in the second decade of the twentieth century, as I learned from Joan Attuat. When we did our last interviews together in 1987, Attuat, almost into her eighties,[3] was living in the growing community of Rankin Inlet, where the Maani Ulujuk School is called after her mother. When we first met, however, in 1983, she lived in the small settlement of Whale Cove, breeding her own dogs, going out daily on her skidoo to tend her fish nets.

Attuat's mother's family were Qaernermiut – inland people from the Baker Lake area – but generations of her forebears had worked for the whalers. "Yes, the family used to follow them quite a bit. They were inland people, but they had to go to work, so they went back and forth. They'd commute with the seasons. The reason they returned inland was because they really were not used to the sea mammals for food. So they used to go back inland because they had to have the inland food."

According to Attuat it was the womenfolk of her family who were responsible for her tribe's becoming known as the Kenepetu, the name by which the Qaernermiut are sometimes erroneously called. Her grandmother Kookoo, or perhaps her great grandmother Silu, with her daughters, was out fishing; then, hearing music, the women went aboard a whaler. "It was raining and Kookoo was wet, and somebody called her over and said, 'Come over here and dance with me,' but she had a packsack she was taking off and said, 'Oh, just a minute! Kenepetu! – I'm all wet!' They nicknamed her Kenepetu, and that way they all became Kenepetu!"

Later, some of her family worked with the RCMP around the time the force established the post at Cape Fullerton – "the first patch of white people" – and Attuat has in her possession a picture of her grandmother Kookoo, a copy of a copyright photograph of "Kookookyouock" by Geraldine Moodie. With such ancestry it is not surprising that Attuat has many stories to tell. One concerns how as a child she went along with relatives to trade with possibly the last vessel ever to winter at Marble Island.

"When you step ashore at Marble Island you have to crawl twice – just two steps. That's what our ancestors had to do; they did it, so we do the same thing. When you first arrive you are told you must crawl,

or otherwise the next year you won't be living.

"When I was a little girl I spent part of a winter at Marble Island because that year a ship was frozen in there in the ice. The captain was Suquortaronik – the harpooner [George Cleveland]. When it was almost winter time and the ice began to freeze, Suquortaronik's ship went to Marble Island. My mother Maani Ulujuk and my grandmother and her husband were all travelling on that ship. There had often been people living there at Marble Island – the place is full of qallunaat graves – and so this year when the ship became caught by the ice, Suquortaronik went in there. My mother and the others had to live on shipboard until the ice was solid and frozen; only then could they get off the ship.

"Because my mother Maani had been so young when I was born, I was sent to live with my great-aunt and her husband. They were like grandparents. I wasn't with my mother on the ship, but that winter when my grandparents went to get supplies – the tea and the things the whalers would trade to the Inuit – I went along too. We travelled over from the mainland by dog-team – three dogs. We were pretty hard up at that time, so we needed supplies from that ship.

"When we arrived we were told to crawl if we didn't want to die. I didn't want to die; I looked at my grandparents and they were crawling, so I did the same thing!

"The ship was anchored in the upper area where the waters are deep. The entrance is just a little inlet,

When the Whalers Were Up North

but inside the inlet the harbour is large. There are small fish there. Our relatives were working on the ship, sewing, doing the dishes, doing other things. They always used to go to eat on the anchored ship, so I went along, too, to feast in the harbour on the ship. There was just one ship; if there were other ships wintering that year, they would have been at Repulse Bay. It must have been a Scottish ship because I still recognize the music they always played. They used to have parties, and one time I saw them having a dance. At that time they used to dance on the rocks – anywhere they could find to dance. This only happened in my childhood; by the time I was growing up there were buildings[4] and they used to hold dances inside.

"My family stayed for a short time only because we had come just to get supplies. Soon we moved on to another camp. My stepgrandparents were really too old, so that people on the ship didn't talk to them much. I hardly remember the supplies, except for the molasses, which I liked very much. I don't really remember what we traded, but my grandmother was a very good sewer so she probably gave them beading work or something of that sort. She sewed things that could be looked at – not really used – souvenirs!

"I've been told two ships sank in that harbour, though I remember seeing only one.[5] I heard from my grandson recently that the white people were examining those ships. They tried to lift them up, but they couldn't do it because of the mud.

"From my ancestors I don't remember hearing of any really terrible happenings. But when I was small I used to be sent away when the adults were discussing because I was always talking too much."

Kookoo (Kookookyouock). Geraldine Moodie, 1904–05.
Museum of Mankind, British Museum, London.

When the Whalers Were Up North

CHAPTER 12

The Wreck of the Seduisante

"There were once two captains who had been to the North and each craved to come back again," related a man telling stories of whaling wrecks. "One man was told by his boss that he was to go. But there were two men both craving to go to the North, two captains who had been to the North before. It was hard for the boss to choose, so they fought a duel and the man who won went North."[1]

In spite of dangers there were always men ready to sail to the North. Dangers there were. Talking about Hudson Bay ice conditions in 1897 for a Canadian government expedition investigating commercial sea-route possibilities, Captain E.B. Fisher, who whaled in the Bay for thirty years, said, "Within my recollection six or eight whaling vessels have been lost up north, two in the strait going in, the *Isabella*, one other, a New London vessel, the *Pioneer*; this last was a

Walrus Hunt at the Floe Edge. Osoochiak Pudlat, Cape Dorset. Etching, 1984.

steamer. Both these vessels were crushed in the ice nip; both were lost just above Big Island on their way in, one in July, the other in August. The other vessels were lost – three at Marble Island parted their chains and went ashore, and the other two or three lost in the Welcome on reefs." Captain John O. Spicer remembered more wrecks, those of the bark *George Henry* and the brig *Pavilion*, both lost in the strait.[2]

There were wrecks throughout the whaling years. I heard fragments of a number of long-ago disasters. Vessels perished by fire, in great storms. Inuit in the Lake Harbour area have half-remembered stories of shipwrecked crews that left by rowboats in the spring – "I don't know where they were going, but I think they were trying to get home." They spoke, too, of a vessel so decrepit that it sank when the Inuit stopped bailing. And Igalook Petaulassie of Cape Dorset heard from her grandmother Nirukatsiak, first of the four wives of the great South Baffin leader Inukjuarjuk, how, perhaps in the 1860s, as a young girl just beginning to live with her husband, she had spent the winter with shipwrecked whalers. "She sewed fur clothing for them. Otherwise they'd freeze." Igalook remembers too that her grandmother was astonished to discover that qallunaat had regular mealtimes. "She wouldn't be hungry, but a meal would be there."[3]

Half-memories of many wrecks are alive in Inuit communities, although details are lost. But Inuit of the eastern Arctic well recall the wreck of the *Seduisante*, a disaster of the final whaling years that involved many South Baffin families.

A name to summon speculation still is that of her owner, Osbert Clare Forsyth-Grant, a Scottish laird's son from Ecclesgreig Castle, near Montrose. Inuit call him Mitsiga – Mr Grant. To some he is a romantic figure, a noble amateur among the free traders; he has also been called a drugstore cowboy. In his short career – it spanned only seven years – he lost two ships, and he died in his thirty-second year with his Scots crew in his second shipwreck on the Hudson Strait in 1911. His Inuit crew and their families survived.[4]

About five years into the new century Grant took over the old shore station at Cape Haven on the tip of the Hall Peninsula, the ancient campsite the Inuit call Singaijaq. This had been one of Spicer's stations, and his New London firm C.A. Williams was already operating there in 1860, when the explorer Charles Francis Hall dropped by and noted the old stone dwellings in which the early Inuit people who didn't use the snowhouse used to live. For many years whalers and Inuit met each other at Singaijaq. The camp

and the nearby waters of Newgumiute Bay (Cyrus Field Bay) drew hunters ready to work for the whalers from the camps of Frobisher Bay, from Cumberland Sound, and also from the Hudson Strait. Cape Dorset people I met were born there, spent childhoods there, in the late nineteenth and early twentieth centuries.

Grant lived at Singaijaq in the station's three cabins in the accepted manner with his Inuit companion Nangiaruk, the wife of Gotilliaktuk, who was Mr Mate, the "first trade man for the whalers." "In those days, even though a woman was married she might be borrowed by somebody else," an informant said with some amusement. "Mitsiga and Nangiaruk had children – they are dead today, but there are descendants."

Mary Ipeelie, a great-granddaughter of Gotilliaktuk and Nangiaruk, says, "I remember lots of photographs of Nangiaruk and Mitsiga together, even framed pictures. As a little girl I used to hear many stories about the whalers. People used to talk about their pilot biscuits and how delicious they were, about how good the maktaaq was. My grandfather Ainiak was Nangiaruk's son. He went overseas with Mr Grant [in 1907]. The reason must have been that Mitsiga was fond of him because he was the son of his woman. So he took his stepson with him when he went to his homeland for a visit. Ainiak liked it there. He used to tell it was not a boring place. He understood a little English, and he learned all kinds of things; he learnt knitting, and years later I learned knitting from my grandfather!"

Grant had his first shipwreck with the ketch *Snow-*

drop in 1908. She was lying at anchor in the Countess of Warwick Sound near the popular camping and hunting site of Minguutuuq, about ninety miles from Singaijaq. On board was the entire Singaijaq camp, about sixty-five people, men, women, and children, many from the Hudson Strait. The ship was blown on to the rocks, but all made it safely to shore. Later, crew member Alex Ritchie told the story on the BBC. "Well, we were all thankful for our escape, but we had nowhere to go for shelter until some sails washed ashore. The first sail to be washed ashore was our spare mainsail, so the Eskimo rigged up a tent for us all. Then the next thing to worry us was that we had no food, till later some of the tinned food started to come ashore after our store tanks burst, for which we were very thankful. The storm died away on the fourth day during our stay ashore. After the tent was erected, the Eskimos made fires for us with pieces of blubber which came ashore when the ship's tanks burst. We got all our clothes dried, and heat for ourselves. Everything that was not in tins was spoiled with the oil and blubber washing through it. The Eskimos were good at lighting a fire. They used flint and steel, and dried moss. They know where to get the dried moss; mixed with a cotton-like weed, one would think it shag tobacco. They made wicks for their lamps with this. We got them ashore, but they kept us alive after we were ashore."[5]

Grant was not deterred. After a hungry and harrowing winter during which his Scottish crew lived and hunted with Inuit families, all were rescued by a passing vessel. Grant went home and brought the French-built topsail schooner *Seduisante* and laid plans for a network of outpost depots, setting up wooden huts at Minguutuuq, site of the shipwreck, and at a point on Grinnel Sound, and telling people that he planned a station at Cape Dorset. But Grant and his crew from Peterhead and Dundee perished in the Hudson Strait off Nottingham Island in September 1911.

Though Grant was on board, the vessel was technically under the command of a captain he had hired. This has always added to difficulties in unravelling accounts of the shipwreck – it is frequently unclear whether references to the captain are to the temporary master or to Grant, the owner.

Mary Ipeelie's husband Arnaitok has often heard stories of the wreck in which the white men lost their lives. "Inuit families with children were along as crew, and they were cruising around the Cape Dorset area when their ship was wrecked at Toojak – Nottingham Island. All the qallunaat died.

"The ship was collecting blubber from walrus. The wreck occurred because the captain was more interested in looking for walrus than watching for shallows. The boss earlier had had a mate who was very terrified to go near the shore although the boss wanted him to go right close. Because the mate was afraid, the boss left that mate down south and chose another man to take his place.

"As soon as they got to Nottingham Island the captain went up in the barrel to see if there were walrus around. They didn't notice the water was shallow because they were looking for walrus.

"When the captain realized the ship was wrecked, he rushed to get most of the equipment into the small boats. They took two boats from the ship and the ammunition, food, and some of the possessions. The captain wanted to go ashore, but he asked the crew and one of them wanted to stay on board. Because one wanted to stay, they all stayed."

The white men sent the Inuit ashore. The Inuit urged the white crew to leave the vessel, too, but the qallunaat did not come. They felt there was no danger since they had a whaleboat attached to the vessel. But after the Inuit left, a wild storm blew up and the whaleboat broke away.[6]

At dawn shots rang out.

Arnaitok Ipeelie says, "I don't know if the shots were a signal for help or a farewell.

"When the night was over, it was obvious all were dead. The Inuit went to look around the shore to see if there were bodies. One man had reached the shore, and people felt he had been alive because his hand was stretched out grasping the snow on the shore.

"The captain had sent the Inuit ashore with supplies, with ammunition and food. The Inuit stayed at Toojak all winter, and then, in spring, they moved on to Cape Dorset. When they were still far off on the water, a hunting party saw them and called, 'How are you?' They knew that families were missing and were anxious about them. The answer came, 'We are well but the qallunaat all are lost; all are dead.' Then those who asked the question called, 'Never mind about the qallunaat. If you are all right, that is all that matters.'

"The Inuit and the qallunaat didn't understand each other too well. Although they were living together, there was not too much understanding of each other's feelings."

Four boats of survivors arrived from Toojak, three wooden whaleboats and a sealskin umiaq constructed during the winter. They carried sixty-one people.[7]

After the wreck, stories circulated that there had been dissension among the white crew. The American explorer Donald B. MacMillan, later rear admiral, who wintered with the schooner *Bowdoin* on the southwest coast of Baffin Island in 1921–22, noted the fact in an account he wrote of his voyage: "Because of the strong wind and sea which arose shortly after their landing, the Eskimos were unable to regain the ship. Just before daylight above the roar of the wind and sea several shots were heard, the result, the Eskimos declare, of trouble which had been brewing among the men for some time, but which I believe to have been signals of distress and an appeal to the Eskimos to make another attempt with the boats … "[8]

Arnaitok Ipeelie often wonders about those final shots: "I like to think they knew they were going and that they were saying 'Farewell!' Because one man had wanted to stay, they all stayed, and when Inuit heard the shots, it seemed like one hit human flesh.

Maybe one of the crew was shot. When they gave up hope, maybe the one who had not wanted to go ashore got shot."

People often ask why Grant and his crew persisted in their decision to stay aboard. But early RCMP special constable James Akavak, in Lake Harbour, had one explanation: "They were afraid of conflict. They knew for sure they were going to be shipwrecked. Maybe they did it for safety."

CHAPTER 13

The Active's Last Voyage, 1912-13

The *Active*'s last voyage in 1912–13 was a fateful one. Alcohol, which some Inuit whalers called "forget-me-quick,"[1] caused tragedy on board, but the same voyage may have saved Inuit lives. Unlike American sailing vessels that frequented the Bay, the *Active* was a steam whaler and generally made only summer cruises. Only once before had the *Active* wintered over. In 1912, however, she set out to stay the year, though not in the customary anchorages of Roes Welcome Sound but southeast in Hudson Bay, in the Ottawa Islands. Her master as usual was Alexander Murray. There had been plans for John Murray to winter nearby with the *Albert*, but a quick freeze-up kept the *Albert* in the Welcome.

The artist Kananginak Pootoogook of Cape Dorset shows his father Pootoogook, Inuit leader in the Cape Dorset area, handing a skin to Chesley Russell, well-known Hudson's Bay Company fur trader and post manager. Coloured-pencil and felt-pen drawing, 1981.

Murray's decision to winter in the Ottawa Islands was an unusual one, but the voyage there was not the first. "Captain Murray of the Scotch Whaler *Active* is said to have taken 500 wolves off the Ottawa Islands in 1908," wrote a Hudson's Bay Company official in a 1911 annual report on the fur trade at Ungava posts. Since the middle of the eighteenth century the company had traded in northern Quebec, first on the east Hudson Bay coast, then in Ungava, opening and closing posts with various degrees of success, but in 1909 it had begun the establishment of posts on the Hudson Strait and set the stage for enhanced trade with the Inuit and the aggressive push into the Arctic lands, in pursuit of white-fox fur, that would characterize the next decades. Perhaps because of Murray's success, and possibly also that of French fur traders, the HBC had sent personnel to winter on the islands, but the venture ended in disaster: "A party of six, two white men and four eskimos sent last year from Port Harrison to the Ottawa Islands to winter there, were all lost in a breeze of wind. It is reported that many bears and walrus are on the Ottawa Islands, also that the Hudson's Bay people do not get half of the fox off these outlying islands."[2]

Perhaps the *Active*'s presence there in 1912 had something to do with the fact that a new era in Arctic history was about to begin. The HBC in 1911 had crossed the Hudson Strait, establishing its first fur-trading post on Baffin Island at Lake Harbour, the point where whaler and Inuit had so often met. According to John Murray's son, Austin Murray of Wormit, Scotland, the two whaler brothers were also planning a new venture. "The idea was that they'd set up their own trading operation, but it was one of those things that was not to be."[3]

Alexander Murray's Inuit son Ikidluak of Lake Harbour was along on what was the *Active*'s last voyage to the Bay. "The second time the ship wintered, I was on it. The Inuit brought their dogs and qamutiit – the sleds – and the dogs were tied on the deck. They spent their time on the boat in the summer, but they camped in winter. We lived in igloos; the ship was not in sight.

"The captain died on that last trip. Before we spent the winter, the captain was already gone. He died probably not from illness but from alcohol, and he was bound up by his crew. Probably the crew was afraid of him. Why they tied him up, I didn't know. Maybe he didn't get along with his crew. I don't know for what reason he was bound. He was shouting out for Inuit people by their names, but the crew would not allow them to go to him. I don't remember hearing whether there was a lot of alcohol on the ship, but I know there was some because the captain died of drink. The crew bound him up."

Alexander Murray, whom the old hunter Kowjakuluk of Lake Harbour described as "a very good man to the Inuit," died 11 November 1912 at a point called Etoile Bay in the Ottawa Islands. The *Active* log confirms that two deaths occurred, that of the captain from "internal tumour" and that of a harpooner from "scurvy." But Robert Flaherty, who met the *Active* in the spring of 1912, reported in *My Eskimo Friends* that two harpooners had died of delirium tremens.[4] During Hudson Bay whaling, alcohol did not cause the debauchery observed after whale ships reached Herschel Island in the western whaling theatre in 1889, but Murray's death can be seen as foreshadowing the family tragedies that alcohol, its presence eventually ubiquitous in the North, was to bring to many Arctic homes. In Dundee, obituaries commented that Murray as a boy had shown great promise, whaling with his father Alexander, the Peterhead master and Franklin search veteran, and succeeding him as captain of the *Windward* at a youthful age. One shipping correspondent reported that the captain had seemed low in mood at the outset of the voyage, hinting perhaps at depression.

But the arrival of the *Active* in the Ottawa Islands was also life-saving. Today in Cape Dorset live a number of families whose members have cause to remember the *Active*'s fateful voyage. During the whaling days the whalers' presence banished to a certain extent the always-present fear of famine. Indeed, I learned that a powerful shaman "could summon a ship if people were starving." That year when the *Active* arrived at her destination in the Ottawa Islands, she let the Inuit off to camp at a site where they found starving people.

Anirnik of Cape Dorset, who was a passenger on the vessel and just becoming a young woman at the time, remembered: "During the year the ship stayed over, the Inuit camped at a place where we found starving people. We caught up with them because the ship left us off there. There were only a very few people left in the camp. The others had starved to death. When the people from the *Active* were walking around, they found those still living in their beds. Those people left from the starvation were brought back on the *Active* with my family. They have offspring here in Cape Dorset today. They were Kenojuak and his wife, Takatak and his wife, and some young people."

One of those young people was Silaki, whom I met in Cape Dorset near the end of her long, eventful life, during which she survived three husbands, delivered many babies (she told me that for her role as midwife she grew her fingernails long to serve her as forceps), and acted as chief mourner when death approached her campmates. She was first pointed out to me as the mother of the artist Kenojuak (called after her grandfather), creator of *The Enchanted Owl*, the most famous of all Eskimo prints.

"During the starvation of which we speak, I don't remember being hungry, but I know that that year people went hungry. My grandmother chewed on ice, and when it melted I drank the water from her mouth; and I got milk from my mother's breast. I don't remember people dying, but I know that that year quite a few people in our camp died.

"The area where we usually camped was a good one, but people get tired of staying in one place, so the year before we had moved to a new camp, crossing over an ice bridge that forms only in certain years; because the bridge didn't form, that's why we starved. We couldn't find animals and we couldn't go back.

"The only reason I think I knew that people were hungry was that my lame father was at the seal hole for a night and a day. The reason my father was lame was that a shaman woman had used her spirit to make his leg useless. We had no dogs but we had a qamutik, and my lame father, Kenojuak, was pulled to the seal hole by my uncle Takatak and other people and he waited there for a night and a day by himself. When he caught a seal, as he was alone, he put wood across the seal hole and tied the seal to the wood. The next day my uncle Takatak and other people came to the seal hole and pulled the seal and my father back.

"My family survived. After they caught that seal we started hunting for foxes. When they caught a fox they would make a fire outside and cook the fox. In those days they didn't have regular traps, so they had to make their own from rocks. They would make a narrow box of rocks with meat of a seal on a string at the end of the box. Attached to the string at the

When the Whalers Were Up North

Sun owl and foliage. Kenojuak Ashevak, Cape Dorset. Lithograph, 1979.

opening of the trap would be a flat rock. When the fox bit the bait, the stone would slam down. They would grab the fox by the feet and kill it by crushing the lungs.

"We were managing quite well when the *Active*

Kenojuak (left) and her sister-in-law Elisapee. They became the wives of Kudlarjuk's sons, Johnniebo Ashevak and Towkie. They are caught by the camera of the Inuit photographer Peter Pitseolak, about 1950. Notman Photographic Archives, McCord Museum of Canadian History, Montreal.

arrived. The ship dropped off the Inuit whalers with a whaleboat – a sailboat with oars, no motor – and left them there till the following year. They spent one year in our camp and caught whales that summer. One time one boat capsized. They were harpooning a whale, and the boat tangled in the rope. All the people were saved; they were rescued by kayak and brought to shore.

"I can't believe the Inuit ever really killed whales on their own. The whale is such a huge creature. They had to have the whalers' equipment. I just can't see them killing a whale with only harpoons. The only way of killing the bowhead whale was with cannons that would explode inside the whale. This time they used the kayaks to carry the harpoons. The kayak didn't carry the cannon; the kayak carried the harpoons and regular rifles. The hunters used the harpoon, and then at the end they fired the cannon [from the whaleboat], making sure the explosive went into the lungs. They used three different kinds of harpoons. They were all handmade, and the weapon part was of walrus tusk.

"I don't remember them catching the whales, but I remember we lived mostly off fresh maktaaq. The stomach part was so soft and thin, different from the beluga maktaaq, which is quite thick and tough. We never ate the flesh of the big whale raw; if it was necessary to eat it, we always cooked it. We used to set fox traps up near the whale carcass, and we'd use the meat and the inside of the whale as bait.

"My brother Tikituk and others were born during

that year. When the *Active* came back to pick us up, my brother, Tikituk, was in the amautiq. I remember being on the *Active* so well. What I remember mostly is seeing so many tents on top of the ship and also inside. That's how the families were living on the ship."

Eventually Silaki's parents reached the hunting grounds around Cape Dorset, where more children were born and where, in the community of Cape Dorset, descendants live today. In time Silaki's own daughter Kenojuak married Johnniebo, son of Kudlarjuk, the little girl who posed with her parents in the New London photography studio at the time of the Case of the Missing Whales, whose family history was also determined, long ago, by the presence of the whalers in the eastern Arctic.

Ikidluak, who told us of Captain Murray's death, supplied more details of the homeward journey. "That year the ship rescued starving people. By the time we arrived, they weren't starving, but during the winter they had gone hungry. They wanted to leave for another land. Takatak, the head of that camp, wanted to move. Takatak and Kenojuak were the only men left of the camp. The rest had starved. Takatak asked the captain of the ship to bring him here to Lake Harbour. He said he could pay with fox skins; he had a few foxes. He told the captain he wanted a gun and a passage for the family.

"We spent the winter down there in the Ottawa Islands and came back to Lake Harbour the next summer. The ship anchored close, and we came to land.

My grandmother picked me up.

"She was Surusimiituq. She was one of the people who lived mostly in the area of Akuliak, and I heard from my grandmother when I was a little boy about how the American ships used to anchor between Akuliak and Iqaqtilik. The Americans were the first to come to this land. People used to say, when they worked for the Americans, they would be paid sooner. I haven't heard that they were paid more. But as soon as the Inuit finished their work, they would be paid. The Scots would pay when they dropped the Inuit off at their camps.

"Most knew when the *Active* dropped us off after that trip when the captain died that the whalers weren't coming back, although maybe some learned only after the ship left. The Hudson's Bay Company already had its post here. It was the year after the Bay built in Lake Harbour that the ship wintered down in the Ottawa Islands. One year after the HBC came here, the ship made its last trip.

"I used to see the weapons from the *Active* for years. Almost every man had one. After the voyage the people on the ship received the weapons they had used. They were long, and the Inuit cut them in two and put wooden handles on them. They got another weapon out of the second half. They cut them to whatever length they wanted so they could be used to kill the harpooned animal. They used them for catching birds, seals. The men had them for years; they kept them for years.

"Yes, the Inuit enjoyed the whaling; they even

speared more whales than the qallunaat. When they were pulling the dead whales, from what I've heard, they used to sing. I've heard the song, though the first words are not too fresh in my mind. My grandmother used to sing the song when she was working around the house. She'd whistle it.

Your good hands; your good feet.
Move them; move them.
You don't mind being wet; you don't mind getting soaked!
Kowk! Blubber!

Tuungaja ... tuungaja ... tuungaja
Good hands! Good feet Ja-gee-ja
Qiirq! Hurry!

"When they said 'Qiirq!' they'd start rowing faster. They sang this song when they were towing a whale. They sang it so as not to lose the whale. That's what they sang when they were pulling the whale to land and they didn't want to lose the whale."

When the Whalers Were Up North

CHAPTER 14

The Last Whales

In their last decades the whalers were traders as much as whalers, and after 1911, when the Hudson's Bay Company built its first Baffin Island post at Lake Harbour and began its sweep into the Arctic territories in pursuit of white-fox furs, the traders in many ways took the whalers' place, inheriting the customs and systems developed in the whaling years. The Inuit bosses – they have been called "powerful like medieval kings"[1] – seem frequently to have retained their positions. Saila, who sailed as a child with Captain John Spicer, became boss for Sikusilaaq, the whole vast Foxe Peninsula where the Hudson's Bay Company established a post at Cape Dorset in 1913.

His son Pauta Saila of Cape Dorset often heard how Saila and his men built an inuksuk to mark the headland of the inlet that the HBC vessel was to enter with the wood for building the post. The inuksuk still

The last bowhead caught in Cumberland Sound. July 1946.
G. Anderson. Hudson's Bay Company Archives, Winnipeg.

stands. "The reason that inuksuk has wood inside is so they could put in some sort of cloth that would flap. My father told them to put a flag there."

In the fall another inuksuk was built. For many years, when the HBC supply ship was expected, hunters would go to this inuksuk with their telescopes and pitch their tents. "When the smoke from the ship was seen, they'd fire their rifles."

Pitseolak Ashoona knew Cape Dorset before a single house was built, and she was there "to run around" when her father helped to build this beacon. "I don't know exactly why this inuksuk was made, but I think it was so that the qallunaat who were coming to Cape Dorset would know their way in. There were several men, and they piled up rocks so high that a man had to get on top so the men at the bottom could hand him up stones. When he held the last big rock, all the stones quivered – but they didn't fall down."

The traders arrived at Chesterfield Inlet on the west coast of Hudson Bay in 1912. "The HBC and the Mission came together," says Father Rolland Courtemanche, who in recent years served with the Oblate mission there. "They came on the same ship. Father Turquetil and one of the company chiefs – maybe Parsons – picked the place. The company had lumber on board and offloaded. The mission brought their house the following year."[2]

As the Hudson's Bay Company marched forward through the Canadian Arctic lands, building post after post, many Inuit moved to hunt and trap for the company in new territory.

The last qallunaat whalers left the Bay in 1915, but the last commercial whaling in Hudson Bay was done for the Hudson's Bay Company, and in Cape Dorset and Lake Harbour live a number of people who as children in the years between 1919 and 1924 watched their fathers kill what were among the last bowhead whales taken in the Bay. Starting in 1919 Sam Ford, HBC post manager at Coats Island, south of Southampton Island, pursued bowhead whales with Cape Dorset and Lake Harbour Inuit for several years, although with only makeshift equipment.

"The company was going after the foxes and polar bears, so our parents were transferred down to Appatuuqjuaq [Coats Island] below Saliit [Southampton Island] so they could get lots of furs," Oola Kiponik of Lake Harbour told me. "People used to go there for caribou hunting. Even though this place is a little island people say the caribou is never scarce. We went on a boat – the *Nanook* – with the Bay manager and Johnny, the interpreter, my family and other Inuit families, and went into this area where nobody lived.

There used to be many whales there. Their colour was almost black, and you could hear them make sounds from under the water. People heard them way under the water."

In the bright spring nights the children kept a whale watch. "We would stay up very late because of the light, and we'd go for walks to a place where we'd watch for whales. Sometimes we saw them, but our parents were asleep," the Lake Harbour hunter Kowjakuluk, who was there with his family, remembers.

All the children of the camp were there that day in 1921[3] to watch as their fathers hunted the bowhead whale. Preparations had been under way for some time. "Ford employed making a bombshell," reads an entry in the HBC Coats Island post journal for 22 May 1920. "Material used viz: 1 lb Powder, 1 pc tin Lard Pail, 1 Thumb screw belong to an old lantern, pc of steel barrel for the point or barb, & 1 pc wood. I have not yet completed it, as I am short of several things. This (bomb or Explosive shell) I intend to use this coming summer on the large Whale that was in this Harbour so often last fall. I am quite sure that we can secure on with it, the only drawback I have no good fuse but I shall try to manage with some of my own make."[4]

One of the young boys watching was Osoochiak Pudlat of Cape Dorset. The missions were well established in the North by the time Osoochiak saw the whale hunt of which he told his daughter, our interpreter Marta Pudlat, and me. As far as he remembers, the shaman's magic played no role. Old customs and responses prevailed long after the missionaries arrived, but the whale hunters he remembers hunted "without superstition."

"When I was a boy there were no shamans as far as I remember. And I haven't heard of them using a shaman to help get the whale. They hunted without thinking something was true when it was not – without superstition.

"When I was a little boy Inuktakaub – the New Person [the missionary A.L. Fleming] – was already living at Lake Harbour. This was before we went to Coats Island and before he became a bishop. He would go to all the little camps to spread the gospel. He never missed a camp. He would travel by dog-team and go to all to teach the word of God. He always came to our family's igloo and built a little igloo as a bedroom with the same entrance as ours. He taught us children Our Father, one hymn, and another short prayer. I am not sure if he taught the same things to adults because he taught the adults separately, but this is what he taught the children. He was a very good man, loving towards the people, and no other bishop will ever compare. The others were good men, but nobody else has been so joyous, so much fun."

Osoochiak drew a delightful, lively picture of Inuktakaub spreading the gospel, and he also drew several pictures of the whale hunt he witnessed. He says, "Yes, I have a lot of things to tell about whalers. When I was a young boy on Appatuuqjuaq – the place of many akpaks [the thick-billed murres of Coats Island]

there were two whale feasts and two whale hunts. It was sometime around the 1920s. We didn't worry then about what year it was or what day. I was just a young boy, ten or eleven, too young to go along, but they got the first whale right offshore, and the whole camp was watching because most people had not seen whalers catch a whale before.

"The whales were inside a little harbour, so you could see almost everything that was going on. In the spring before the ice broke up, the whales always came there to feed under the ice on the little fish with the big eyes. You could see them spraying through the cracks in the ice. That was their feeding area. The whales would sit on the floor of the ocean or the top of the sea and let the brown hair fringe at the ends of the baleen float out and gather in those tiny fish with big eyes.

"The whalers were Suu, Ningiutsiaq, Sitaa, Sapanqaq, Paujungi, Melia, Kuutuu, Qalipallik. They didn't hunt for themselves; they were lent the whaling gun and the equipment for catching whales. If they did catch a whale, they'd be given a lot of things by the HBC.

"The whalers had two boats – a whaleboat and an

OVER:

The young missionary A.L. Fleming spreading the Gospel. In syllabic writing is a prayer Fleming taught his congregations. Osoochiak Pudlat, Cape Dorset. Coloured-pencil and felt-pen drawing, 1981.

When the Whalers Were Up North

The Last Whales   153

ordinary wooden boat. On the whaleboat they had the equipment for catching the whale. Though I didn't see them use it, the whaling gun they had was like a little cannon that shot short fat bullets; the harpoon and the line were shot at the same time. It could kill instantly.

"When the whaling ships were up here, there were always several whaleboats – one with a gun to shoot the whale, another with a gun which would shoot a harpoon with rope attached. There'd be another with a barrel with ropes made of skin, and two guys worked with the ropes – one would make sure the two ropes didn't get tangled, and the other would be constantly pouring water on the ropes to keep the rope running freely when it caught on to the whale. The rope would steam from the heat because the friction was so great from the speed. They had a board across the barrel to prevent the rope flying out all at once.

"When the whales are wounded they go down and stay down for a very long time – maybe half an hour – and then come up here or there.

"When they shot the whale at Appatuuqjuaq that day there was rope on the harpoon. The only reason they caught the whale that day was that it was a very calm day; they were rowing all over the place."

According to the old Lake Harbour hunter Kowjak-uluk, the whale hunters had a close call. "Those whalers nearly had an accident. When they were nearing the arvik, the steer man went too close to the fin, and the whale fought back. The seat and the boom of the mast broke on the impact."

Osoochiak noted that on another whale hunt the hunters used their kayaks. "The second whale they caught that year was caught by men in kayaks. They took the harpoon gun on the whaleboat, and then when the whale was in pain the men went after it on kayaks and used their guns. The second whale was bigger than the first.

"In the old days they caught their whales with only homemade harpoons. There used to be a lot of them caught around Southampton and Coats Island – the proof is that there are so many whale bones all over the place. The hunters would go right alongside the whale, close enough to stick in the harpoon. But it wasn't frightening at all, even on a kayak, because after you hit, the whale breathes twice before it goes down. And bowheads don't seem to react as quickly as other animals when wounded.

"When the whaling ships came up here, there'd be a lot of Inuit on the ships. A whaling ship [the *Seduisante*] sank near Toojak [Nottingham Island] the year I was born. There used to be a lot of whaling around here [South Baffin Island] and around Pang-nirtung. A great deal around Pangnirtung. I'm not sure if the qallunaat were happy or unhappy when they were around the Inuit camps, but most likely they were fitting in and enjoying themselves. I've heard a few stories from the people who were on the ships, and they always said they were taken care of very well and transportation for hunting was provided. The whalers wanted polar-bear skins and sealskins, so they'd make the hunting easier for the Inuit with them. They wanted bearskins, sealskins, and the baleen. That's what they took back with them. Never any meat, just skins and baleen. The ships around here [in the later whaling days] weren't after the whales so much; they were after the skins.

"The older Inuit folk liked working with the whalers, but the young folk were just a bit afraid of the whales. There was one young fellow who was manoeuvring the boat, but when the moment came, he didn't turn it properly because he was afraid to go right up to the whale.

"They always had a feast after the killing of the whale. The whale I watched being caught that day was captured in the late spring, and they had to wait until the tide was out to cut it up. They had special instruments with big long handles attached to a knife. They took off their clothes and put on oilcloth slickers – the colour used to be yellow, although today they have different colours. They still wore their kamiit – sealskin boots – but they received the oilcloth clothing free from the HBC because they were hunting for whales. They put it on because the blubber would get all over their skin clothing. They cut slits along the side of the whale in order to get to the top. You could squeeze your foot in and climb that way.

"When the whalers started packing the fat, I helped, too. They used forty-five-gallon drums to pack the fat, and it added up to a lot of forty-five-gallon drums. As the whalers were cutting up, they were usually eating, and after they finished we gathered together and ate some more.[5]

"Those animals were so big, as high as the ceiling. The whale I saw caught that day they said was only a young one, but to me it looked like a giant whale. I know that whales were caught so huge that if you put the kayak upright it would reach to the top of the flipper.

"That's how they used to measure a big whale."

In the decades after the whale hunts off Coats Island, from time to time there were whales caught in Hudson Bay; they were caught as the first whales had been by the Inuit, for their own use. When the whaling days were almost over, the Inuit whaling crews who had worked for Comer began around 1908 to resettle Southampton Island, deserted after the last Tuniit died, and eventually Shoofly's son John Ell became chief of a new Sadlimiut people. Comer himself surveyed Southampton Island and found the good anchorage he called Coral Harbour – because of a coral-like material brought up there when taking soundings – where island residents live today.[6] In the decades after the Americans and Scots departed, Inuit whalers from Coral Harbour (and also Repulse Bay) caught several bowheads. The hunters used equip-

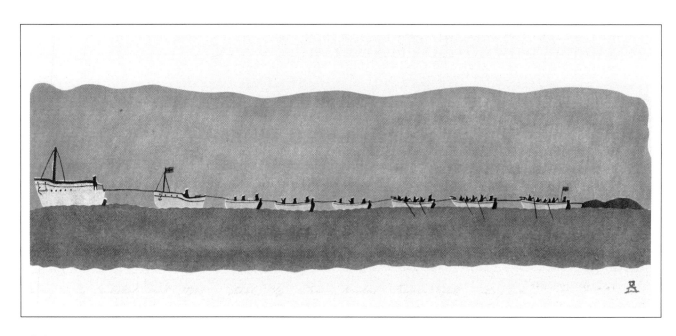

Whaling in Cumberland Sound, 1930s. Tommy Nuvaqirq,
Pangnirtung. Stencil, 1977. Although the 1930s is the date
given for the whale hunt shown here, it seems likely that the
print depicts the successful whale hunt of July 1946.

ment left behind by the foreign whalers (they say there is still qallunaat whaling equipment in Inuit possession), but they caught the whales for their own purposes, sometimes selling part of the whale products to the Hudson's Bay Company.

One whale was caught by John Ell. His daughter Ookpik remembers that John Ell came back from hunting with a flag flying from the mast. "It might have been a Union Jack – that meant something was caught."

Sandy Santiana, the adopted son of John Ell, was along on the expedition. "I was with my father twice when we caught whales. I was about fifteen. I was a little bit scared, but I had to keep concentrating on

The HBC *Motor boat* Ungava *returning to Pongnirtuing with
dead whale in tow, July 1946.*
George Anderson, Hudson's Bay Archives, Winnipeg.

what I was doing. John Ell and I were in one boat,
and there was one other boat along. We had an idea
where the whale might be, but we had to find him.
We had no motor, so we were running him down by
sail. Every time the whale surfaced, we'd sail to that
spot. We used the harpoon gun, and the harpoon
caught on to him, but the explosive didn't go off.
When the whale started pulling the boat, you could

see the water rushing by. The whale moved so fast that the boat went right up, and we could see the whale beneath us behind the boat. We weren't tied to the whale, so we could have let go if we had become scared. We put in another cartridge and fired again, and this time the whale slowed right down.

"We towed the whale to land, and it was hard work cutting the whale up. It took us a few days, but somebody had to do it; we just kept at it. We left only a little meat that had spoiled on the carcass. We kept the blubber and some of the oil for eating, and the meat we kept for the dogs. We put most everything into barrels or cached it under rocks. The baleen we sold. The carcass attracted all the polar bears and wildlife, so that was a year when we had good hunting. The maktaaq was the best I've ever had. It's the best of all the whales."

Santiana and his father hunted one more whale and killed it. They got a little of the meat, but the weather was dark and stormy, so they had to let it go. "After that we never looked for whales again."

But Santiana still has a memento of these whaling expeditions. In the old days – indeed, from time immemorial – the runners of the Inuit sleds were made from the whale's jaws. "In my boat I keep a piece of the sled that was made from the jaw of the whale my father caught."

Today the remaining bowheads swim with little disturbance. In the waters of Cumberland Sound people say that the whales are coming back. Ross Peyton, who owns the Clearwater Fishing Lodge at the head of the sound, says a bowhead often observed swimming nearby recently appeared with a calf. The last whale caught by the sound's Inuit whalers was taken in 1946, some twenty years after regular whale hunts ceased. It was caught by Angmalik, in the days of the stations the boss at Kekerten. Elisapee Ishulutaq, one of Pangnirtung's popular graphic artists, was born in the 1920s and is too young to recall the whaling days, but she does recall, most vividly, how young and old, from camps all along the sound, travelled to see Angmalik's bowhead – the greatest quarry of them all, pursuit of which directed the lives and fortunes of her people for one hundred years. "Word spread," she says, "that Angmalik had caught a whale, so we rushed from our camp over to see. It was in an inlet north of Pangnirtung. It was quite an exciting time. I was not a child. I was a young woman, and I stood beside it and looked way up."

Koodlu Pitseolak, photographed on Blacklead Island at about age twenty with her baby daughter Makiktuq up on her back. Koodlu's husband Marcusie was one of the last Cumberland Sound whalers. In 1989 she was living in Pangnirtung.
About 1922–23.

From the album of Captain Edmund Mack. Notman Photographic Archives,
McCord Museum of Canadian History, Montreal.

When the Whalers Were Up North

Final Curtain

In retrospect many dramatic and sometimes tragic events seem to presage and symbolize the ending of the whaling days. In Cumberland Sound the *Ernest William*, after wintering so many years around Repulse Bay, was wrecked at Kekerten in 1913. Her anchor can still be seen there. Her log for Thursday, 4 September, reads: "Vessel became a total wreck on the Rocks on Kikkerton Island."[1]

Etooangat Aksayook, only "this high" at the time and too young to use a rifle, witnessed the event. "The ship had become so very dangerous she could not be taken home. So she was anchored in front of Kekerten and the cargo taken off, and then she was beached and left on land for the Inuit to salvage." The vessel's six crew members, now facing the prospect of a long winter at Kekerten, remembered that the previous day they had seen another vessel down

Children Dancing to the Whaler's Jig. Elisapee Ishulutaq, Iqaluit, formerly Pangnirtung. Stencil, 1983.

the shore. As it happened, that vessel's captain was Jimmy Mutch, who had run the Kekerten station for many years prior to setting up an operation near Pond Inlet at Albert Harbour a few years after the turn of the century. Etooangat says, "The crew, who had no way to go home, headed up the lookout hill with a big drum of fuel. In the evening they lit a big bonfire. It was calm on the water and they knew the ship would be moving slowly, so they kept the fire going. When it was almost morning, the ship arrived and two people came up. It was then that I saw Jimmy Mutch – all the old workers, the old men, the old women recognized him and were so happy. 'My old boss,' they kept saying. Jimmy had not expected to come by that way. He hugged a lot of them and just stood there looking."

People said that since Jimmy had been the boss they had all grown old.

The *Tilly* sank in the sound in 1916, and the *Easonian*, owned by Kinnes's Cumberland Gulf Trading Company Ltd., of Dundee, the vessel that the Inuit of Cumberland Sound call the last whaler, came to a spectacular end at Kekerten in 1922. Etooangat, now a hunter, had been along with the whaleboats on her last journey. "We were whaling and on our way back to Kekerten when the engines failed. From there on they had to use sails. When we reached Kekerten, it was planned to beach her when the tide went out, and we started tugging the ship with the whaleboats to shore. Then there were cries, 'She's on fire!' It started in the grease around the engines. It started just like that! The engines weren't working, but some sort of spark ignited the grease. The Inuit kept pulling the ship by means of the long rope taken from whaleboat to whaleboat. By this time the ship was really ablaze. We got some of the cargo off and the Inuit kept on trying to tow the ship until the fire broke out to the outside. She burned right in the harbour facing Kekerten."

In Hudson Bay there were also disasters and tragedies. Harry, Comer's whaling boss, lost many members of his family crew in a dreadful accident in the shifting ice in the dangerous currents off Wager Bay about 1913. Uvinik, Harry's adopted son, lost both his wives, and later Uvinik shot himself in his grief. Captain Alexander Murray died ice-bound in the Ottawa Islands in 1912, a quick freeze-up having prevented his brother John from joining him with the *Albert* to winter nearby.[2] John Murray took the *Active* home, laden with six whales, one of the last great catches of the whaling years.[3] He ventured out with her one more time, but the doughty old vessel was

shortly sunk, a casualty of the First World War. Comer's last vessel, the *A.T. Gifford*, made the last whaling voyage to the Bay under Captain Gibbons, but the *Gifford* sank in flames in 1915 on the homeward journey, with all hands aboard. Captain Comer returned to the Bay once more, in 1919, on what was to be an exploring and trading expedition. But disaster dogged even this veteran of the Bay; he lost his vessel the *Finback* in the waters off Cape Fullerton. "It happened just because I was foolish enough to move on the high tide, which I didn't realize until after I came out of the harbour. As I got out to seaward, everything was covered with tide. In trying to avoid one reef, the first thing I knew I was on another. There was no one to blame but myself. And I probably knew that harbour better than anyone else."[4]

These retired whalers and their families, mostly living on Southampton Island, were photographed with their parish priest Father Lionel Ducharme OMI in Chesterfield Inlet in 1923. Bernadette Ookpik Patterk, second from the left, bottom row, and her son John identified the people in this picture. Top row, left to right: Shoofly (Nivisinark, and also called Susangna), Scotch Tom (Angutimmarik), By n' By (Siattiaq), Helen Paniruluk, Eugene Qigley, Panikutuapik with baby Paapaq. Bottom row: Tommy Bruce (Tom Luce, also known as Ukpaqtuq), Bernadette Ookpik Patterk, Igviksaq, Father Lionel Ducharme, Anulik, Ungalaaq, Laurent Pameolik, Nicodine Quasaq.
Archives Deschâtelets, Oblate de Marie-Immaculée, Ottawa.

Comer set out in a whaleboat and sailed to Chesterfield Inlet, where the Hudson's Bay Company supply boat *Nascopie* made annual calls. Everyone who could, Inuit from near and far, would be in Chesterfield for shiptime.

Perhaps it is the stuff of legend, but Eugenie Tautoonie Kablutok, then a bright-eyed little girl, believes she once caught a glimpse at Chesterfield of both Captain Comer and the Scottish whaling captain John Murray. "They were old men then and they were on the ship that was going back and forth until it sank [probably the *Nascopie*, wrecked off Cape Dorset in 1947]. I remember my parents saying, 'There is Cross Eyes; there is Angakkuq.' I remember that very well. Nakungajuq – Cross Eyes – probably wanted to see his girlfriend and bring her presents. He had had this girlfriend for so long; he must have loved her very much. Angakkuq came back because he wanted to see John Ell and his girlfriend Nivisinaaq.

"In the whaling days these captains had their own people, so after the whaling was done, they came back one more time to see their old helpers, to see their girlfriends, and to check on their own people."

Murray does not appear to have been in the Bay in 1919, but he continued to voyage in the Arctic long after the whaling days were over and was in command of the HBC supply vessel *Nascopie* from 1928 to 1930. Comer's last appearance in Hudson Bay was on the *Finback*, wrecked on the reefs at Fullerton on 27 August, 1919.

Shortly after his last voyage, Captain Comer left on record his own comment on the interaction between the whalers and the Aivilik. "We got attached to the natives ... " he said.[5] "They have worked among the whalers since Civil War times, and coming in contact with them has changed the natives ... While whalers perhaps have not always been what they should be, still I have found them just as good as any other people I ever met."

During my talks I often asked informants, descendants of the Aivilik, for their impressions of the impact of the whaling years. They gave reasoned responses.

Whalers brought liquor, informants explained, but they drank "not too much, only a bit." They brought guns, but ammunition sometimes ran out. Upon occasion Inuit whaled so late in the fall that the usual preparations for winter were neglected.

"One of the main benefits was that the Inuit gained some weapons from these people," Joe Curley said. "But once the ammunition was all used up, they were not very useful. And it would be November before the whaling stopped. Sometimes there were starvations, because people couldn't survive if they did not have the caribou clothing they needed for the winter months. People working for the whalers sometimes had no time to go out and hunt for caribou during the shedding season – the time when you catch the caribou before the coats are too thick. We used to see people walking around with real thick clothing, and it made them quite uncomfortable. Some of them looked very shabby. They used to be short sometimes of warm clothing and also of the caribou hides that used to be needed for mattresses. This happened not really *because* of the whalers but because the Inuit travelled so much with them. The men did not go out hunting to get the caribou mats and clothing – the mitts and all the garments."

But whalers had their virtues. They did not seek to dominate in the way Inuit sometimes feel present-day qallunaat do. "They didn't establish power bases," a young interpreter said. Harry Kilabuk of Iqaluit, who lived at the east Baffin Singaijaq whaling station in its last days, elaborated: "In the whaling days the qallunaaq just hopped on the sled without paying for the trip or the hunter asking for anything. The qallunaaq used to get a ride and eat, and the Inuit never thought about pay. They were living in harmony together. The qallunaat lived just like the Inuit, and they used to help each other out. Right now we are starting to get into these communities. Before we lived in different places – anywhere we wanted. But when the Inuit were working for the whalers, I have heard people say, even though we were beginning to be run by white people, they were really happy, and enjoyed what they were doing."

And Joe Curley struck a similar note: "I really can't recall any disagreements between the Inuit and the qallunaat. Sometimes the whalers would have the mate do something he didn't want to do, but there were

John Murray aboard the Hudson's Bay Company supply ship the Nascopie, *1928–30.*   *Courtesy of Austin Murray.*

never feelings of inferiority or superiority between whalers and Inuit. They would help each other out as best they could. It must have been a long winter for those people to put in on their ships. Most of their arguments would occur because of the long winters; they longed for home.

"They were pretty good among the Inuit people." Long after the whaling days were done, the influence of the American and Scottish whalers continued and, like the wild sweet accordion music they introduced, still echoes in Inuit life.

The story that perhaps best provides a final curtain for the whaling days concerns the old whaling station of Singaijaq. After the wreck of the *Seduisante*, in which Mitsiga – Mr Grant – and many qallunaat lost their lives, Singaijaq passed into other hands and for a time continued its life as a station. Then in 1923 its owners, like other free traders, sold out to the Hudson's Bay Company. For a while the campers remained, but changing times, brought about by the company's pursuit of white-fox fur, soon forced them to move to better trapping.

But for years after the whaling station at Singaijaq finally stood abandoned, Inuit regularly returned to Singaijaq to visit – and sometimes felt ghosts around them. Mary Ipeelie of Iqaluit recalls: "There were three houses built at Singaijaq. Mitsiga – Mr Grant – lived in one, one was for storage, and the third was also for storage but for not important storage. I'm not sure who originally built these cabins, but they were built probably well over one hundred years ago. They say two white men lived in them for years.

"I myself never saw people living in those cabins, but quite a few times when I was a child we stopped there on our way caribou hunting.

"We used to hold dances in one of the cabins. There were accordions stored there that belonged to the Inuit. They'd been left in the cabin to keep them safe, so they'd last longer. There were thirty or more, perhaps thirty-seven or thirty-eight. The people who owned them didn't dare take them in case they'd get wrecked.

"We danced by the light of oil lanterns. In those days we used to call dancing mumiq – turning opposite – but now most people just say dancing – it's more common. There were so many accordions – some more elaborate than others – and the Inuit were so happy to dance. Everyone took turns playing, my mother included.

"I think Mitsiga – Mr Grant – may have been the one from whom people got their accordions.

"We kept on dancing in those cabins for years. Every summer when we went caribou hunting we passed by the cabins and danced.

"I can tell you a story about the dances and about something I saw when I was five years old which will really interest you. Suddenly, while they were having a big dance, all the lights went out, and people thought there was a polar bear outside. I remember my father grabbed me and took me to our tent. That night the dogs kept on barking all night long. I remember so well the sound of the barking. There were doors banging and strange noises. People felt something was going on. Next day there was talk about Mitsiga and the men who had died in the wreck. People thought perhaps they did not want dances in the cabin. But nothing like that ever happened again. It was just the one time; although we held dances again, nothing particular ever happened.

"So they continued dancing for years. Around 1939 they realized the whalers were not coming back, and they began to take the lumber from the cabins. The Inuit used it for tent frames and for the floors of the tents. Finally my mother also took her accordions away. There were two – one remained good, but we children wrecked the other by taking out the different parts.

"I remember those cabins so well. There were dishes and household things still stored in them and blubber oil stored in the barrels. No one touched them for years, thinking the qallunaat would come back."

When the Whalers Were Up North

It seems appropriate to outline briefly the approaches taken in collecting and presenting the oral history around which this book is built. While some interviews were collected earlier and the last were collected in 1988, this book first took shape during the most active period of interviewing, while I was retained by the Urgent Ethnology Programme of the Canadian Museum of Civilization in 1982 and 1983 to collect interviews in communities on the west coast of Hudson Bay and on South Baffin Island, partly, according to my contract, because of "journalistic" experience. In Canada public institutions such as the Canadian Museum of Civilization have been the main proponents of oral history, while in the United States universities took the lead, with Columbia University establishing the first oral history program in the country in 1948.[1] Under its auspices American journalists published some of the first contemporary oral history. In 1977 Bernard Ostry, at the time secretary-general of the National Museums of Canada, said, "There is a new urgency to the work of recording the thoughts and feelings and values of people and communities who have been left too long in the dark and who are now suffering not only the insults of history but the shock of technical and social change. And in this urgent endeavour the National Museums of Canada are playing their part ... The task is to offer a deeper

Measuring the Whale's Tail. Osoochiak Pudlat, Cape Dorset. Stonecut, 1986.

orchestration to the elegant formulations of academic history, not to supplant it but to augment it by something new in the world, a secular history."[2] As a journalist participating in the collection of interviews for the Canadian Museum of Civilization, I considered it my task to collect the material available, to present it fairly and accurately, and to make it accessible to the reader by providing adequate historical context. My aim has been to present Inuit reminiscences so they stand as a resource and a record, but not to attempt interpretation or analysis as a geographer or historian might do.

Over almost a decade a total of fifty-two people were interviewed, with forty-five supplying information that appears here. Thus, the great majority of informants passed on information of interest, while a number have been major contributors. All interviews were tape-recorded, and in eliciting information I relied on a reporter's mainstays: question and answer. Informants, who were paid a research fee, were customarily interviewed at least twice, some intensively over a week or more. A few informants like Anirnik, Pitseolak Ashoona, Pauta Saila, and Osuitok Ipeelee of Cape Dorset had accorded me interviews while I worked on Canada Council grants on earlier visits to the North, and stories that took on particular pertinence in the light of the present study have been extracted from these earlier interviews. In 1987, thanks to the Multiculturalism Program of the Department of Secretary of State, I was able to collect additional interviews in Iqaluit and Pangnirtung and to revisit

Joan Attuat in Rankin Inlet and Osoochiak Pudlat and Osuitok Ipeelee in Cape Dorset. In 1988 an invitation from Nunatta Sunaqutangit, Iqaluit's "Museum of Things from the Land," permitted me to visit Pond Inlet.

In placing material from the transcripts into the text, in the interests of readability I have transposed paragraphs, sometimes sentences, and assembled continuous narrative from sections of transcript. In doing so, I have made every effort never to distort words or intentions. Tapes and transcripts of the interviews collected for the Urgent Ethnology Programme are deposited at the Canadian Museum of Civilization in Ottawa.

I would like to express here my admiration for the interpreters of the North. In the early 1970s Ann Meekitjuk Hanson, now deputy commissioner of the Northwest Territories, who began her career as an interpreter, told me she felt her role was to "help people understand." I received such help from many talented individuals. With the exception of a few short sections, all tapes were interpreted twice, once at the time of the interview, then again as soon as possible thereafter, when the interview would be typed up so that a further interview could supply additional detail. Whenever possible, two interpreters worked on each interview, one participating in the interview, the second supplying the important back-up interpretation. Interpretation of the tapes was done as carefully as conditions in the field permitted; a number of translations of particularly difficult tapes or puzzling passages were made later in the south by Susan Gardener Black.

Expert interpreters have many demands on their time and can expect appropriate recompense, so I was fortunate indeed that in the Hudson Bay area the Northwest Territorial Interpreters' Corps generously provided many of the necessary second interpretations. In Iqaluit, David Audlakiak, an executive with Bell Canada, was kind enough to supply interpretation for particularly difficult sections of a south Baffin Island tape.

While the intention was always to allow Inuit voices to speak for themselves, it was obviously necessary to supply a historical framework. In providing this I have tried wherever possible to make use of excerpts from early published sources and from logbooks, journals, and documents of the period. Perhaps it is worth mentioning that when collecting oral history in the North it is often difficult for an interviewer to establish an appropriate time-frame for an informant's reminiscences. I have always found informants to be most anxious that their information should be true, and some found it preferable to withhold information for which they did not have all the details. However, it is not always easy to know the years in which the incidents related took place. Very few older Inuit have exact knowledge of their date of birth, and estimates can be out by many years. In the early 1970s I sat in on a meeting in Cape Dorset called specifically so that people could try to determine their ages, which were often roughly arrived at by relating a person's birth to a significant event: "I was born the summer after the company came" – that is, a year after the Hudson's Bay Company established its trading post in Cape Dorset (1913). There is a tendency sometimes to assume the old are older than they are; to the young the old are very old – "My children say I am eighty" – and reaching old age with its reward of the old-age pension may sometimes have been a desirable goal. Fortunately, establishment of the time-frame can often be achieved by means of the so-called paper trail. Surprisingly often, an excellent trail exists. It was sometimes possible to document informants' stories both through early manuscript or published sources and photographically, as in the Case of the Missing Whales. Here a photograph of the Inuit involved in the case existed, and legal documents provided passages that both bore out and supplemented Inuit accounts. Informants were usually asked if they could make drawings to illustrate their stories, so sometimes we have their own graphic documentation as well.

A word on the spelling of Inuit words is in order. The spelling of Inuit words in roman orthography has been in flux for many years, and a consensus is only now emerging. Deborah Evaluarjuk, a linguistic specialist with Indian and Northern Affairs Canada, kindly checked the spellings of many Inuit words in this manuscript to help achieve conformity. Inuit words that have entered the English language – igloo, kayak – have been given their accepted English spelling. In the case of proper names, where a particular spelling has been widely used, I have followed precedent. I have also tried to employ the spelling preferred for their names by informants and interpreters. Inuit adopted surnames widely only in the early 1970s (one or two of the oldest informants had no surnames), but even where the same surname has been adopted by a number of family members, variation in spelling has sometimes developed.

Of those interviewed, five persons – Anirnik of Cape Dorset, Joe Curley of Eskimo Point, Leah Arnaujaq of Repulse Bay, Leah Nutaraq of Iqaluit, and Etooangat Aksayook of Pangnirtung – had vivid firsthand impressions of the last of the whaling days and were also sometimes able to pass on second- and third-hand stories from much earlier times. A number of informants knew the whalers as small children or sailed as infants up on their mothers' backs on the last of the whaling voyages and were able to pass on some first- as well as second-hand material. Several informants were able to give eyewitness testimony to whale catches that occurred after whaling proper ended. Second-hand reminiscences came from elders who grew up in families that had once had close involvement with the whalers: "I missed the whaling days, but I knew the Inuit whalers," one man explained. These informants often passed on stories from their personal family traditions. Three who provided valuable information functioned as community historians – Kowjakuluk of Lake Harbour, Cornelius Nutarak of Pond Inlet, and Eugenie Tautoonie Kablutok of Rankin Inlet – and had done their own research on the whaling days. During visits to communities I did occasionally meet persons who preferred not to be interviewed. But this was unusual. Most people I approached seemed to feel it was important to preserve a record of the past. "It is all I can do nowadays – give information," Eugenie Tautoonie Kablutok remarked. And Agee Temela of Lake Harbour, after I thanked her for her interview, said "It is a pleasure to do something worthwhile." Today, many whose voices were recorded on the tapes are no longer with us. Joe Curley died within a year of the interviews we did together in 1983: he had received one of the two tags the community gives out for hunting polar bear and died living Inuit life to the full, on a polar-bear hunt. Today there are fewer and fewer to tell the stories.

For the most part archival photographs and contemporary Inuit graphics illustrate this book. The drawings and prints capture old memories; in contrast, the photographs captured the moment before it turned to history. Some of the Inuit images were drawn by informants to illustrate aspects of our talks; a number are drawings, stencil or stonecut prints selected from material held by Eskimo co-operatives in the Inuit communities and from public and private collections of Inuit art. In the case of stencils and stonecut prints, the artists' images, created first on paper, have been transferred to the print medium and executed by highly skilled Inuit specialists. Co-operatives usually issue annual editions of prints. Copyright to most Inuit works reproduced here is retained by the artists through their co-operatives: the West Baffin Eskimo Co-operative, Cape Dorset; the Pangnirtung Eskimo Co-operative; and the Sanavik Co-operative of Baker Lake. The archival photographs come from repositories in Canada, the United States, England, and Scotland. The photographers observed Inuit society in the late nineteenth and early twentieth centuries and sometimes saw the Inuit whalers at work.

INFORMANTS

LAKE HARBOUR: Agee Temela, Oola Kiponik, Kowjakuluk, Isaccie Ikidluak, James Akavak

CAPE DORSET: Pauta Saila, Pitaloosie Saila, Anirnik Oshuitoq, Pitseolak Ashoona, Peter Pitseolak, Osoochiak Pudlat, Pudlo Pudlat, Osuitok Ipeelee, Igalook Petaulassie, Kiawak Ashoona, Ashevak Ezeesiak, Kenojuak Ashevak, Tye Adla, Kalai Adla, Mary Qayuaryuk, Egevadluk Ragee, Ulayu Pingwartok, Keeleemeeoomee Samualie, Silaki

IQALUIT (FROBISHER BAY): Harry Kilabuk, Mary Ipeelie, Arnaitok Ipeelie, Leah Nutaraq, Ben Ell, Elisapee Ishulutaq, Alookie Ishullutaq

PANGNIRTUNG: Etooangat Aksayook

POND INLET: Cornelius Nutarak, Simon Anaviapik

ESKIMO POINT: Joe Curley, Jackie Napayok

RANKIN INLET: Ookpik Patterk, Robert Tatty, Sandy Santiana, Eugenie Tautoonie Kablutok, Joan Attuat

CHESTERFIELD INLET: Maria Teresa Krako, Leonie Egalak Sammurtok

WHALE COVE: Joe Uluksit, Eve Alakasuak

BAKER LAKE: Tommy Tapatai, Ruth Annaqtussi

REPULSE BAY: Leah Arnaujaq, Rosa Kanayuk

CORAL HARBOUR: Kanajuq Bruce, Samson Ell, Mary Ningeogan

INTERPRETERS

LAKE HARBOUR: Bea Ikidluak, Johnny Manning, Jallie Akavak, Sam Pitseolak, Mary Akavak, Peesee Pitseolak, Gila Pitseolak, Pitseolak Akavak

CAPE DORSET: Pauta Saila, Pitaloosie Saila, Anirnik Oshuitoq, Pitseolak Ashoona, Peter Pitseolak, Osoochiak Pudlat, Pudlo Pudlat, Osuitok Ipeelee, Igalook Petaulassie, Kiawak Ashoona, Ashevak Ezeesiak, Kenojuak Ashevak, Tye Adla, Kalai Adla, Mary Qayuaryuk, Egevadluk Ragee, Ulayu Pingwartok, Keeleemeeoomee Samualie, Silaki

PANGNIRTUNG: Meeka Wilson

POND INLET: Carmen Kayak

ESKIMO POINT: Madelaine Napayok Anderson

RANKIN INLET: Rosie Aggark, John Patterk, Luke Issaluke, Sally Tatty, Elizabeth Lyall, Jacob Partridge, Mikle Langenhan

WHALE COVE: Mary Jane Ford

BAKER LAKE: Joe Mautaritnaaq

REPULSE BAY: Steve Kopak

CORAL HARBOUR: Sarah Eetuk, Bernadette Panniak Dean

OTTAWA: Susan Gardener Black

NOTES

INTRODUCTION

1 Melville, *Moby Dick*, chap. 6.
2 Therkel Mathiassen's somewhat negative account of the experience appears in *Report on the Expedition*, chap. 6.
3 Leah Arnaujak, Repulse Bay, and Etooangat Aksayook, Pangnirtung, in *Recollections of Inuit Elders*, 9–20, 29–34; Ittuangat Aksaarjuk [Etooangat Aksayook], in "Whaling Days," 21–9.
4 Egevadluk Ragee, Cape Dorset.

PROLOGUE: ARRIVAL OF THE WHALERS

1 Francis, *Arctic Chase*, 23.
2 Leah Arnaujaq, Repulse Bay, who heard the story in her childhood from an elderly woman originally from the East Baffin coast.

CHAPTER 1: IN CUMBERLAND SOUND

1 Boas, *The Central Eskimo*, 59. Nutaraq notes that at Blacklead Island the trader rang a bell at biscuit time; Alookie Ishullutaq of Iqaluit, who lived at Kekerten, confirms that there he blew a horn.
2 Wakeham, *Report of the Expedition to Hudson Bay and Cumberland Gulf in the Steamship "Diana"*, 74.
3 Peter Pitseolak, Cape Dorset. A similar story is also recounted by Etooangat Aksayook of Pangnirtung.
4 Ross, *Arctic Whalers, Icy Seas*, 151.
5 Ibid., 155–73.
6 Wakeham, *Report of the Expedition to Hudson Bay and Cumberland Gulf in the Steamship "Diana"*, 74–5.
7 Francis, *Arctic Chase*, 13.
8 Philip F. Purrington, The Whaling Museum, New Bedford, Massachusetts, personal communication.
9 Etooangat Aksayook, Pangnirtung.
10 Apparently the *Tilly*, according to the log of the *Erme*, which shows the vessel picked up four shipwrecked men from that schooner at Singaijaq 15 Sept. 1916. Stefansson Collection, Dartmouth College.
11 A census conducted on Baffin Island for the Department of the Interior, Northwest Territories and Yukon, appears to locate Nutaraq at Blacklead Island in the winter of 1923–24 and to provide what the census report terms "fair approximation" of her age and those of her children. According to information supplied by Nutaraq, at the conclusion of the whaling era she was already twice widowed and the mother of two children, Markoosie and Leah. Among the family units at Blacklead Island, the census report lists the widow Nootasa, aged 26; a boy Makose, aged 5, and a female child Lea, aged 3. NAC RG 85, vol. 64, file 164–1, pt 1.
12 The HBC bought Kekerten station with its stock and stripped it of trade goods and equipment over a couple of years; it operated Blacklead as an outpost until September 1931.

CHAPTER 2: IN HUDSON BAY AND THE STRAIT

1 Royal North West Mounted Police Sessional Paper no. 28, 6–7, Edward VII, A1907, Appendix N, "Patrol Report, Constable L.E. Seller, Fullerton to Lyons Inlet," 122.
2 Ross, *Whaling and Eskimos*, 37.
3 Pitseolak and Eber, *People from Our Side*, 131.
4 The suggestion is that the shaman would have powerful amulets to provide protection behind him; personal communication from Bernadette Driscoll.
5 "Customs and Traditions: Stories Ancient and Modern Written during the Winters of 10, 11, 12 by George Comer, Master Whaling Schooner *A.T. Gifford*, Cape Fullerton Hudson Bay," ms in Boas collection, American Museum of Natural History, New York.
6 Nourse, ed., *Narrative of the Second Arctic Expedition Made by Charles F. Hall*, 139.
7 Eugenie Tautoonie Kablutok, Rankin Inlet.
8 Joe Curley, Eskimo Point.
9 Gilder, *Schwatka's Search*, 304–5.
10 Zerubavel, *The Seven Day Circle*, 5–26.
11 Ross, *Whaling and Eskimos*, 78.
12 Bernadette Ookpik Patterk, Rankin Inlet.
13 Leah Arnaujaq, Repulse Bay.
14 Pitseolak and Eber, *People from Our Side*, 57.
15 Joe Curley, Eskimo Point.

16 Silaki, Cape Dorset.
17 In a deposition entered in the case of *C.A. Williams et al.* v *Jonathan Bourne et al.*, the journalist Wm. H. Gilder, who accompanied Frederick Schwatka on his 1878 expedition, was asked, "Do they sometimes kill whales with native boats and native weapons?" He answered, "Yes."

CHAPTER 3: SHAMANS AND WHALERS

1 Eleeshushe Parr, Cape Dorset.
2 Osoochiak Pudlat, Cape Dorset.
3 Taylor, "The Arctic Whale Cult in Labrador," *Etudes / Inuit / Studies*, 121–32. Taylor quotes Kaj Birket-Smith, *The Eskimo*, trans. W.E. Calvert (London: Methuen 1936).
4 Mary Ningeogan, Coral Harbour.
5 Pitseolak and Eber, *People from Our Side*, 27.
6 Boas, "Customs and Traditions."
7 From the sketchbook of an unknown whaleman on the *Orray Taft*, log 276–a, Collections of the Kendall Whaling Museum, Sharon, Mass.

CHAPTER 4: THE CASE OF THE MISSING WHALES

1 Hall, *Life with the Esquimaux*, 28–9.
2 Parts of the court records for *C.A. Williams et al.* v *Jonathan Bourne et al.*, Nos. 1653, 1654, before the US Circuit Court for Massachusetts, May 1882, are retained in the Federal Archives and Records Center, Waltham, Mass. Depositions presented as evidence in the cases by Lieut. Frederick Schwatka and Wm H. Gilder are in the G.W. Blunt White Library, Mystic Seaport Museum, Mystic, Conn.
3 Wakeham, *Report of the Expedition to Hudson Bay and Cumberland Gulf in the Steamship "Diana,"* 59.
4 Hall, *Life with the Esquimaux*, vol. 1, 103.
5 *Era* log, June 1879 – Nov. 1880, International Marine Archives, The Whaling Museum, New Bedford, Mass.
6 Rear Admiral Donald MacMillan photographed Peuliak during his 1921–22 voyage with the *Bowdoin* to the Foxe Peninsula, and his unpublished manuscript of the voyage (Special Collections, Bowdoin College Library, Brunswick, Maine) car-

ries mention of a woman who had lived for several winters on Spicer's vessel (possibly on, or in camps nearby, the *Roswell King*, which for a time served as a year-round station based in Spicer's Harbour.)
7 Pitseolak and Eber, *People from Our Side*, 131.
8 Kowjakuluk, Lake Harbour.
9 Gilder, *Schwatka's Search*, 275.
10 Deposition of Frederick Schwatka, G.W. Blunt White Library.
11 Deposition of Wm. Gilder, G.W. Blunt White Library.
12 Court records, Federal Archives and Records Center, Waltham, Mass.
13 R.B. Wall, "Famous Groton skipper had many strange experiences in the frozen north," in supplement to *The Spicer Genealogy*, material originally published in 1921 articles in *New London Evening Day*, New London, Conn.
14 Collections of the New London County Historical Society, New London, Conn.
15 Aggeok Pitseolak, Cape Dorset.
16 The *Mystic Press*, 17 Nov. 1881.
17 Hall, *Life with the Esquimaux*, vol. 1, 103. Hall notes that at the time he met him, John Bull had as a wife Annie Kimilu's sister Kokerzhun.
18 Fleming, *Archibald the Arctic*, 153.
19 Pitseolak and Eber, *People from Our Side*, 131.
20 Oola Kiponik, Lake Harbour.
21 Many details of the case are to be found in "Eskimo Joe and a Point of Law," *Bulletin* (Winter 1959).
22 Wall, *New London Evening Day*, 28 July 1921.
23 A series of five cases, the two most important being *Cherokee Nation* v *State of Georgia* and *Worcester* v *State of Georgia*.
24 Barnard L. Colby, *The Day*, New London, Conn. 9 March 1935.
25 Wall, supplement in *The Spicer Genealogy*.
26 *Roswell King* log 1886–89, International Marine Archives, The Whaling Museum, New Bedford, Mass.
27 Ashevak Ezeesiak, a grandson of Kudlarjuk.
28 Fleming, *Archibald the Arctic*, 154.
29 Isaccie Ikidluak, Lake Harbour.
30 Kowjakuluk, Lake Harbour.
31 Personal communication from the Spicer family.

CHAPTER 5: SPICER'S HARBOUR

1 A map in Wakeham, *"Report of the Expedition to Hudson Bay and Cumberland Gulf in the Steamship "Diana,"* designates some miles of coastline as Spicer territory.
2 Log of the *Era* from June 1879 to November 1880, International Marine Archives, The Whaling Museum, New Bedford, Mass.
3 Ross, *Whaling and Eskimos*, 51–2.
4 Joe Curley, Eskimo Point.

CHAPTER 6: THE SIIKATSI AND THE WRECK OF THE *Polar Star*

1 Inuit pronunciation of "Mr Grant." Osbert Forsyth-Grant was owner of the ill-fated *Snowdrop*, lost in 1908, and *Seduisante*, lost in 1911. See chap. 13.
2 For background information on the free traders I am much indebted to an exchange of correspondence with Gavin White, Department of Ecclesiastical History, University of Glasgow.
3 Wakeham, *Report of the Expedition to Hudson Bay and Cumberland Gulf in the Steamship "Diana,"* 56.
4 Ibid., 55.
5 Ibid., 75.
6 Personal communication from the Spicer family.
7 Currently in possession of Austin Murray of Wormit, Scotland.
8 Lake Harbour and Cape Dorset people consider that kayaks disappeared on the coast in the 1940s.
9 Isaccie Ikidluak, Lake Harbour.

CHAPTER 7: DEATH OF THE LAST TUNIIT

1 Ross, *Whaling and Eskimos*, 116.
2 Eleeshushe Parr, Cape Dorset.
3 Dr L.D. Livingstone, 6 March 1925, to O.S. Finnie, director, Northwest Territories and Yukon Department of the Interior, Ottawa, NAC RG 85/815, file 6954.
4 Pitseolak and Eber, *People from Our Side*, 33.

5 Ross, *Whaling and Eskimos*, 116.
6 Ferguson, *Arctic Harpooner*, 216.
7 Pitseolak and Eber, *People from Our Side*, 31–5.
8 Comer, "Description of Southampton Island and Notes upon the Eskimo," 86–9.
9 Ross, *Whaling and Eskimos*, 116.
10 Pitseolak and Eber, *People from Our Side*, 31–5.
11 Mary Ningeogan, Coral Harbour.
12 Therkel Mathiassen, *Archaeology of the Central Eskimos*, 284.

CHAPTER 8: THE *Active*

1 Robert Kinnes's journal, in the possession of Robert Kinnes and Sons, Dundee, Scotland.
2 Kinnes clippings, in the possession of Robert Kinnes and Sons, Dundee, Scotland.
3 1912–13 in the Ottawa Islands.

CHAPTER 9: WINTERING

1 Barrie Kinnes, personal communication.

CHAPTER 10: PERSONALITIES OF THE BAY

1 The Oblate missionaries brought Christianity to the west coast of Hudson Bay in 1912, when they established their first base at Chesterfield Inlet just prior to the final whaling voyages out of the Bay. Anglican missionaries had arrived earlier on Baffin Island, setting up a mission in the Cumberland Sound in 1894 and a Hudson Strait mission station in 1909 at Lake Harbour, close to where the Robert Kinnes Company of Dundee had a shore station. But the shamans ceded their authority slowly; for many years shamans and missionaries coexisted.
2 Ross, ed., *An Arctic Whaling Diary*, 92.
3 Manuscript journal of George Comer on board the *Era* 1897–99, G.E. Blunt White Library, Mystic Seaport Museum, Mystic, Conn.
4 Ross, ed., *An Arctic Whaling Diary*, 74.
5 Saladin d'Anglure, "Les masques de Boas," 168.

6 Spaeth, *Read 'Em and Weep*, 63; Dichter and Shapiro, *Handbook of Early American Sheet Music*, 150–1; Shapiro and Pollock, *Popular Music*, 1608.

7 Samson Ell, Coral Harbour.

8 Ross, ed., *An Arctic Whaling Diary*, 192.

9 Ibid., 191.

10 Ibid., 192.

11 George Diveky, personal communication.

12 Driscoll, *The Inuit Amoutik*, 17–18.

13 Ross, ed., *An Arctic Whaling Diary*, 95.

14 Joan Eldridge, letter to the author, 15 Nov. 1984.

15 Joan Eldridge, letter to Hugh Dempsey. Glenbow Museum, Calgary, Alta., 6 Aug. 1985.

16 Geraldine Moodie was more than an amateur. In a letter dated 30 Aug. 1985 to Joan Eldridge, Georgeen Klassen, assistant chief archivist of the Glenbow Museum, writes, "Our information on Mrs. Moodie's photography career shows that she learned the craft in the mid-1890s while her husband was stationed at Battleford. By the time he was transferred to Maple Creek in 1896 she was operating a studio in Battleford and in 1897 opened studios in Maple Creek and Medicine Hat." Because of Canadian copyright regulations at the time, a particularly useful collection of Geraldine Moodie photographs is held by the Ethnology Department of the British Museum, London. In this collection the photographer has identified her subjects by their Inuit names, which has made possible the collection of information relating to a number of the people photographed. The phonetic spelling of the often difficult-to-pronounce Inuit names – the reason for the frequent bestowal of the so-called whaling names – does, however, often present problems.

17 Ross, ed., *An Arctic Whaling Diary*, 151–2.

18 Joan Eldridge, personal communication.

19 Ross, ed., *An Arctic Whaling Diary*, 151–3.

20 *Report of the Royal North West Mounted Police*, Sessional Papers 28 A 1905 to 28 A 1916 (Ottawa: King's Printer A 1910), 258.

21 Joan Attuat, Rankin Inlet.

22 Ibid.

23 Driscoll, "Sapangat: Inuit Beadwork in the Canadian Arctic," 46.

24 George Comer to Vilhajalmur Stefansson, 20 September 1927, Stefansson Collection, Dartmouth College Library, Hanover NH.

25 Ben Ell's biological father was Pameolik, one of the children adopted by Shoofly and said to be Comer's natural son. According to Joan Attuat of Rankin Inlet, who spent her infancy in the same family, Pameolik lived with his mother, a Qaernermiut or Kenepetu woman by the name of Ookok or Ooktok, but after her death was adopted by Shoofly. According to Eugenie Tautoonie Kablutok of Rankin Inlet, the daughter by another partner of John Ell's first wife, John Ell was Comer's son by Shoofly. "Everybody at the time knew they had a child together." But this has been disputed. "I've heard both," says Ben Ell. "I've heard he was; I've heard he wasn't."

26 Marie Patterson, Iqaluit.

27 Silaki, Cape Dorset.

28 Quoted in "The Arctic Socialite from Beetlebung Road," *Bulletin*, (Summer 1963): 1–40.

29 Ibid., and *New Bedford Standard*, 12 Nov. 1905.

30 Ross, "George Comer, Franz Boas, and the American Museum of Natural History," 160.

CHAPTER 11: THE LAST VESSEL AT MARBLE ISLAND

1 Journal of the bark *Orray Taft*, vol. 276, Kendall Whaling Museum, Sharon, Mass.

2 Sketch notebook of an unknown whaleman on the bark *Orray Taft*, log 276a, Kendall Whaling Museum, Sharon, Mass.

3 Joan Attuat believes that her father may have been a man called Walker with the RCMP at Cape Fullerton. A record of births compiled by whaling master George Comer while in the Cape Fullerton area notes, "Policeman Walkers Child a Girl born the fall of 1909 Mother Known as Marney Kenepetu," typescript list in the Comer papers, Mystic Seaport Museum, Mystic, Conn.

4 Dances were often held in the HBC warehouses.

5 Some uncertainty existed as to the location of the wrecks of the whalers *Orray Taft* and *Ansel Gibbs* until recent investiga-

tions indicated that one vessel sank in the harbour and one outside the gut. However, William H. Gilder, in 1881 in *Schwatka's Search*, noted that the hulk of the *Orray Taft* "lay upon the shore of the inner harbor, but the *Ansel Gibbs* broke up outside and had long since gone to pieces" (37).

CHAPTER 12: THE WRECK OF THE *Seduisante*

1 Arnaitok Ipeelie, Iqaluit (Frobisher Bay).
2 Wakeham, *Report of the Expedition to Hudson Bay and Cumberland Gulf in the Steamship "Diana,"* 56–8.
3 Nirukatsiak lived, Igalook thought, into the 1920s. "I am an old woman now," she said, "but my grandmother when I remember her was much older." The events she related may have taken place in the 1860s. Inukjuarjuk, Nirukatsiak's husband, had had an earlier, well-documented experience with the white man: see Francis and Morantz, *Partners in Furs*, 140–1, and Pitseolak and Eber, *People from Our Side*, 21–6. As a boy in Arctic Quebec, Inukjuarjuk had been along when he and his brothers were forced by a jealous faction in their camp to kill shipwrecked white men for their metal goods and possessions. He was tattooed across the bridge of the nose because he had killed qallunaat. In those days, his son Peter Pitseolak reported, "They used to tattoo the men when they got something big." The Baffin Island encounter with shipwrecked white men Nirukatsiak reported as entirely friendly.
4 For a detailed treatment of Grant's career, see Fraser and Rannie, *Arctic Adventure*.
5 Fraser and Rannie, *Arctic Adventure*, 111.
6 Pitseolak and Eber, *People from Our Side*, 79–83.
7 Ibid., 83.
8 Unpublished ms by Rear Admiral Donald B. MacMillan, Special Collections, Bowdoin College Library, Brunswick, Maine.

CHAPTER 13: THE *Active*'S LAST VOYAGE

1 Ross, *Whaling and Eskimos*, 124.
2 H.M.S. Cotter in "Report on Fur Trade – Labrador District, for year ending 31 May 1911," in Supplementary Report on

Ungava Posts, Hudson's Bay Company Archives, Provincial Archives of Manitoba, A74/20, f 62.
3 Personal communication.
4 Ross, *Whaling and Eskimos*, 124.

CHAPTER 14: THE LAST WHALES

1 James Houston, personal communication.
2 Father Arsene Turquetil, OMI later bishop; and Ralph Parsons, who became the last fur-trade commissioner for the Hudson's Bay Company.
3 Reeves, Mitchell, Mansfield, McLaughlin, "Distribution and Migration of the Bowhead Whale, *Balaena mysticetus*, in the Eastern North American Arctic," 32.
4 Entry in Coats Island post journal, 22 May 1920, Hudson's Bay Company Archives, Provincial Archives of Manitoba, Winnipeg, B404/a/1, f 74.
5 Kowjakuluk of Lake Harbour also remembers the Appatuuqjuaq whale feast: "Yes, the maktaaq was very very tasty, much better than the beluga. The meat was never eaten because the bomb harpoons exploded inside the whale. When the whale breathed after it was hit, it would breathe out black soot. As children we were told, 'Never touch the meat.' Because the meat was no longer good. When they pulled the whale to land that day, I remember just before they reached the shore, three men hopped on the floating creature and had their pictures taken."
6 In recent years many descendants of the Inuit whalers who made Southampton their home have moved southward down the coast to communities more in the mainstream of modern-day Inuit life.

CHAPTER 15: FINAL CURTAIN

1 Log of the *Ernest William*, 1913 voyage, Dundee Public Library, Dundee, Scotland.
2 *Buchan Observer and East Aberdeenshire Advertiser*, Peterhead, Scotland, 7 Oct. 1913.
3 Ibid., 21 Oct. 1913.
4 "Interview with Captain George Comer," 21 Oct. 1919, ts,

Stefansson Collection, Dartmouth College Library, Hanover,
NH.
5 Ibid.

A NOTE ON THE TEXT

1 Peter Stursberg, "Oral History," *The Canadian Encyclopedia*
(Edmonton: Hurtig 1985), 1331.
2 Ostry, Bernard, "The Illusion of Understanding: Making the
Ambiguous Intelligible," address to the 11th National Collo-
quium on Oral History, Sept. 1976, in Montebello, PQ; later
printed in *Canadian Oral History Journal*, Mar. 1977.

Aksaarjuk, Ittuangat [Etooangat Aksayook]. "Whaling Days: My Life with a Whaling Crew." *Isumasi – Your Thoughts* (Eskimo Point, NWT: Inuit Cultural Institute) 1, no 2 (Oct. 1987): 21–9.

Anderson, George. "A Whale Is Killed." *Beaver* (Mar. 1947): 18–21.

Boas, Franz. *The Central Eskimo.* Washington: Bureau of Ethnology 1888.

Bockstoce, John R. *Whales, Ice and Men: The History of Whaling in the Western Arctic.* Seattle: University of Washington Press and New Bedford Whaling Museum, Massachusetts, 1986.

Bulletin [New Bedford, Mass.]. Old Dartmouth Historical Society and Whaling Museum, Johnny Cake Hill, New Bedford, Mass. "Eskimo Joe and a Point of Law" (Winter 1959): 1–4; and "The Arctic Socialite from Beetlebung Road" (Summer 1963): 1–4.

Calabretta, Fred. "Captain George Comer and the Arctic." *The Log of Mystic Seaport* 35, no. 4 (Winter 1984): 118–31.

Catalogues of the Pangnirtung Inuit Co-op 1973–88.

Catalogues of the West Baffin Eskimo Co-op 1959–88.

Clark, G.V. "An Arctic Veteran." *Beaver* (Summer 1970): 64–7.

Colby, Barnard L. *The Day* (New London, Conn.), 9 March 1935.

Comer, George. "A Geographical Description of Southampton Island and Notes upon the Eskimo." *Bulletin of the American Geographical Society* 42 (Feb. 1910): 86–9.

Dichter, Harry, and Elliot Shapiro. *Handbook of Early American Sheet Music 1768–1889.* New York: Dover 1977.

Douglas, W.O. "Two Whales within the Hour." *North* (July-Aug. 1975): 8–13.

— "The Wreck of the 'Finback.' " *Beaver* (Spring 1977): 16–21.

Driscoll, Bernadette. *The Inuit Amoutik: I Like My Hood To Be Full.* Winnipeg: Winnipeg Art Gallery 1980.

— "Sapangat: Inuit Beadwork in the Canadian Arctic." In *Expedition: The University of Pennsylvania Museum Magazine of Archaeology / Anthropology* 26, no. 2 (Winter 1984).

Eber, Dorothy Harley. "Eskimo Penny Fashions." *North* (Jan.–Feb. 1973): 37–9.

— "On Koodjuk's Trail: Robert Flaherty's photographs evoke the past for Cape Dorset Eskimo." *Natural History* (Jan. 1979): 78–85.

— "Bringing the Captain Back to the Bay: Photographs taken by a nineteenth-century whaling captain evoke memories among present-day Inuit." *Natural History* (Jan. 1985): 66–73.

Ferguson, Robert. *Arctic Harpooner: A voyage on the Schooner Abbie Bradford 1878–79.* Ed. Leslie Dalrymple Stair. Philadelphia: University of Philadelphia Press 1938.

Fleming, Archibald Lang. *Archibald the Arctic.* New York: Appleton-Century-Crofts, Inc. 1956.

Francis, Daniel. *Arctic Chase: A History of Whaling in Canada's North.* St. John's, Newfoundland: Breakwater Books 1984.

Francis, Daniel, and Toby Morantz. *Partners in Furs: A History of the Fur Trade in Eastern James Bay 1600–1870.* Kingston and Montreal: McGill-Queen's University Press, 1985.

Fraser, Robert J., and William F. Rannie. *Arctic Adventurer: Grant and the Sedusisante.* Lincoln, Ont.: W.F. Rannie 1972.

Gilder, William H. *Schwatka's Search: Sledging in the Arctic in Quest of the Franklin Records.* New York: Charles Scribner's Sons 1881.

Goldring, Philip. "Inuit Economic Responses to Euro-American Contacts: Southeast Baffin Island, 1824–1940." Canadian Historical Association, *Historical Papers 1986*, 146–72.

— "Religion, Missions, and Native Culture." *Journal of the Canadian Church Historical Society* 26, no. 2: 43–9.

Hall, C.F. *Life with the Esquimaux: The Narrative of Captain Charles Francis Hall, of the Whaling Barque "George Henry," from the 29th May, 1860, to the 13th September, 1862.* 2 vols. London: Sampson Low, Son and Marston 1864.

Harper, Ken. "Pangnirtung." *Inuktitut* (Spring 1985): 18–36.

Lipton, Barbara. *Arctic Vision: Art of the Canadian Inuit.* Ottawa: Canadian Arctic Producers and Indian and Northern Affairs Canada, 1984.

Low, A.P. *Cruise of the Neptune: Report on the Dominion Government Expedition to Hudson Bay and the Arctic Islands on board the DGS Neptune.* Ottawa: Government Printing Bureau 1906.

Mathiassen. Therkel. *Archaeology of the Central Eskimos.* Vol. 4, Report of the Fifth Thule Expedition, 1921–24. Copenhagen: Gyldendal 1927.

— *Report on the Expedition.* Vol. 1, Report of the Fifth Thule Expedition, 1921–24. Copenhagen: Glydendal 1945.

Mitchell, Edward, and Randall R. Reeves. "Catch History and Cumulative Catch Estimates of Initial Population Size of Cetaceans in the Eastern Canadian Arctic." Thirty-first report of the International Whaling Commission, Cambridge, England, 645–82.

Morrison, William R. *Showing the Flag: The Mounted Police and Canadian Sovereignty in the North 1894–1925*. Vancouver: University of British Columbia Press 1985.

Nouse, J.E. ed. *Narrative of the Second Arctic Expedition Made by Charles F. Hall 1864–69*. Washington: Government Printing Office 1879.

Ostry, Bernard. "The Illusion of Understanding: Making the Ambiguous Intelligible." *Canadian Oral History Journal* (Mar. 1977).

Pitseolak, Peter, and Dorothy Eber. *People from Our Side*. Edmonton: Hurtig Publishers 1975.

Recollections of Inuit Elders: In the Days of the Whalers and Other Stories. Inuit Autobiography Series No. 2. Eskimo Point, NWT: Inuit Cultural Institute 1986.

Reeves, Randall, Edward Mitchell, Arthur Mansfield, and Michele McLaughlin. "Distribution and Migration of the Bowhead Whale, *Balaena mysticetus*, in the Eastern North American Arctic." *Arctic* 36, no. 1 (Mar. 1983): 5–64.

Ross, W. Gillies, *Whaling and Eskimos: Hudson Bay 1860–1915*. Publications in Ethnology No. 10. Ottawa: National Museum of Man 1975.

— ed. *An Arctic Whaling Diary: The Journal of Captain George Comer in Hudson Bay 1903–1905*. Toronto: University of Toronto Press 1984.

— "George Comer, Franz Boas, and the American Museum of Natural History." *Etudes / Inuit / Studies* 8, no. 1 (1984): 145–64.

— *Arctic Whalers Icy Seas: Narratives of the Davis Strait Whale Fishery*. Toronto: Irwin Publishing 1985.

Saladin d'Anglure, B. "Les masques de Boas: Franz Boas et l'ethnographie des Inuit." *Études / Inuit / Studies* 8, no. 1 (1984): 165–79.

Shapiro, Nat, and Bruce Pollock. *Popular Music: A Revised Cumulation*. Vol. 2. Detroit: Gale Research Company 1986.

Spaeth, Sigmund. *Read 'Em and Weep: The Songs You Forgot To Remember*. New York: Doubleday, Page & Company 1927.

Spence, Bill. *Harpooned: The Story of Whaling*. Greenwich, England: Conway Maritime Press 1980.

Spicer Genealogy. Privately published family papers in possession of John Spicer, Groton, Conn.

Stackpole, Renny A. *American Whaling in Hudson Bay 1861–1919*. Mystic, Conn.: Munson Institute of American Maritime History, The Marine Historical Association, Incorporated 1969.

Taylor, J. Garth. "The Arctic Whale Cult in Labrador." *Études / Inuit / Studies* 9, no. 2 (1985): 121–32.

Wakeham, William. *Report of the Expedition to Hudson Bay and Cumberland Gulf in the Steamship "Diana" under the Command of William Wakeham in the year 1897*. Ottawa: Queen's Printer 1898.

Wall, R.B. "Famous Groton skipper had many strange experiences in the frozen north." Supp. to *The Spicer Genealogy*, material originally published in 1921 in articles in the *New London Evening Day*.

White, Gavin. "Scottish Traders to Baffin Island 1910–1930." *Maritime History* 5, no. 1 (Spring 1977).

Zerubavel, Eviatar. *The Seven Day Circle*. New York: MacMillan 1985.

Bibliography

INDEX